"The Voyage of
the *F.H. Moore*"
and Other 19th Century
Whaling Accounts

"The Voyage of the *F.H. Moore*" and Other 19th Century Whaling Accounts

SAMUEL GRANT WILLIAMS,
J. ROSS BROWNE,
CAPT. CHARLES H. ROBBINS
and FRANCIS ALLYN OLMSTED

Edited by Greg Bailey

McFarland & Company, Inc., Publishers
Jefferson, North Carolina

LIBRARY OF CONGRESS CATALOGUING-IN-PUBLICATION DATA

Williams, Samuel Grant, 1852?–
　"The voyage of the F.H. Moore" and other 19th century whaling accounts / Samuel Grant Williams, J. Ross Browne, Capt. Charles H. Robbins, and Francis Allyn Olmsted ; edited by Greg Bailey.
　　p.　　cm.
　Includes index.

　ISBN 978-0-7864-7866-8 (softcover : acid free paper) ∞
　ISBN 978-1-4766-1368-0 (ebook)

　1. F.H. Moore (Brig)　2. Whaling—History—19th century.
3. Whaling—Massachusetts—History.　4. Seafaring life—History—19th century.　5. Whaling.　I. Browne, J. Ross (John Ross), 1821–1875.　II. Robbins, Charles Henry.　III. Olmsted, Francis Allyn, 1819–1844.　IV. Bailey, Greg, 1954–　V. Title.
SH381.G64 2014
639.2'8092—dc23　　　　　　　　　　　　　　2014010989

BRITISH LIBRARY CATALOGUING DATA ARE AVAILABLE

© 2014 Greg Bailey. All rights reserved

No part of this book may be reproduced or transmitted in any form or by any means, electronic or mechanical, including photocopying or recording, or by any information storage and retrieval system, without permission in writing from the publisher.

On the cover: Whaling scene © 2014 Pictures Now; background images iStockphoto/Thinkstock

Printed in the United States of America

McFarland & Company, Inc., Publishers
　Box 611, Jefferson, North Carolina 28640
　　www.mcfarlandpub.com

To my parents—GB

"[B]e a Columbus to whole new continents and worlds within you, opening new channels, not of trade, but of thought."
>From *Walden*
By Henry David Thoreau

Table of Contents

Preface — 1

**PART I: "THE VOYAGE OF THE *F.H. MOORE*"
BY SAMUEL GRANT WILLIAMS** — 5

One. Meet Fluker and the Crew 5
Two. Our First Day at Sea: Goodbye to Boston 9
Three. On the Way to San Domingo 13
Four. Story of Captain Peakes 17
Five. Gonives Bay and Our First Whale 22
Six. Off for More Whales 27
Seven. Off This Coast of Honduras 31
Eight. Blackfish and More Fruit 33
Nine. Cruising in the Tracks of Columbus 36
Ten. Cruising Among the Islands 38
Eleven. Southwest Key and One Comfortable Home 40
Twelve. New Year's, 1874 41
Thirteen. Sam the Boatsteerer's Story 45
Fourteen. [Untitled] 48
Fifteen. The Captain's Wife 58
Sixteen. [Untitled] 62
Conclusion 69

**PART II: AN UNFINISHED BOOK
BY SAMUEL GRANT WILLIAMS** — 71

One 73
Two 75

Table of Contents

Three 77
Four 79
[An Incomplete Chapter] 81
Canary Islands 86

**PART III: EXCERPT FROM *INCIDENTS OF A WHALING VOYAGE*
BY FRANCIS ALLYN OLMSTED — 89**

**PART IV: EXCERPTS FROM *ETCHINGS OF A WHALING CRUISE*
BY J. ROSS BROWNE — 99**

**PART V: EXCERPTS FROM *THE GAM*
BY CAPT. CHARLES H. ROBBINS — 163**

Making a Master 164
That Great Leviathan 167
Beach Combers 177
Right Whales 180
Whale-Land and Its Customs 182
The Frozen North 186
Whales Has Feelins 189
The Captain 197

Index — 203

Preface

Before the invention of the light bulb in 1879, the world at night was a dark place. Only the dim light of fires, candles and lamps fueled by whale oil or, increasingly, kerosene made from coal, penetrated the darkness that dominated the hours from dusk until dawn. The rhythms of everyday life were ruled by natural light cycles—almost inconceivable today with our twenty-four-hour lifestyles—and the cleaning, refilling and lighting of oil lamps was a part of a daily routine. In order to function after sundown, even in a limited way, people of the 19th century needed a ready source of fuel for their imperfect lamps. With the petroleum industry in its infancy, the most available and desirable lamp fuel was whale oil—also known as "train oil"—obtained by the industrial-scale hunting, killing and processing of whales.

Today, the practice of whaling is widely viewed with abhorrence. But in the 19th century, the taking of whales was necessary to daily life, and not a subject of moral debate or social criticism. In addition to furnishing premium lamp fuel, whale oil was also rendered into soap, and whale bone—the plastic of the 1800s—was made into corsets, knife handles, combs, umbrella spokes, collar stays, piano keys and countless other household items. Whales were simply a commodity, and their harvesting was as guilt-free as the catching of any fish from the ocean. Like the bison which once covered the North American landscape to the horizon, whales seemed an inexhaustible natural resource to the people of the 19th century; any notion that, like the bison, whales were being hunted almost to the point of extinction had not yet disturbed the public conscience.

At its peak in the mid–19th century, whaling was the fifth largest industry in America, employing thousands of men on hundreds of ships operating across the world's oceans. But by the spring of 1873, the American whaling industry was in decline. The Civil War had all but put an end to

Preface

whaling, as the Confederate navy targeted the New England–based whaling fleet, and the Union navy expropriated a number of whalers, filling them with stones and sinking them in Charleston Harbor in an unsuccessful attempt to blockade the waterway. In what would later be called the Whaling Disaster of 1871, thirty-three New England whaling ships were lost off the northern coast of Alaska, where they became trapped in pack ice and were abandoned. The greatest threat to the industry, though, still loomed: one year earlier, the oil town of Titusville, Pennsylvania, where the petroleum industry was born in 1859, became the first city to pipe in natural gas for illumination.

One of the whaling ships taking to sea in 1873 was the *F.H. Moore*, a two-masted ship called a brig. The *Moore* was one of hundreds of similar ships in the diminishing American whaling fleet after the war. Built in Chelsea, Massachusetts, in April 1868, the *Moore* displaced 107 tons. This would be the ship's fourth recorded voyage.

One eager crew member, Sam Williams, signed on as a boatsteerer, whose job was to direct one of the two small whaleboats carried by the ship and, when in range, to throw harpoons into the whale. Samuel Grant Williams was born in 1852 in Peabody, Massachusetts, but called Portsmouth, New Hampshire, his home. His Welsh grandfather served on a British ship in the War of 1812, but after being captured by the Americans decided to marry and settle in the new land. On his way back to England to settle his affairs, he died of yellow fever in New Orleans. His only child, the father of Sam Williams, was apparently drawn to the sea as well: in the 1860 census, he was not listed with his wife and children. One of the few surviving artifacts of Williams' life is a children's book about a reluctant schoolboy that his mother gave him for Christmas in 1865.

Williams, who had just turned twenty-one, kept a personal log on his voyage, and years later reworked it into the narrative that follows. The voyage turned out to be eventful, even fun at times, but it was the task of Williams and his shipmates to hunt and kill whales, and process their blubber into oil. This they did in ways which may seem barbaric and cruel today, but they were men of their times doing what they were expected to do.

The ship's master, Captain Robert Soper of Provincetown, Massachusetts, was an experienced and wealthy captain, and a veteran of seventeen previous voyages dating back to 1840. He had previously been in the banking business. In 1834, he built an octagonal house in Provincetown which still stands today. In 1861, Soper put out to sea on the whaling ship *Mermaid*

Preface

along with two other ships from Provincetown. They were at sea when the Confederates fired on Ft. Sumter, and had no idea that the Civil War had started when, on May 24, the ships were captured by the rebel navy. Relieved of his ship, Soper and the other captains were taken to New Orleans. After a brief stay in one of best hotels in the city, they were allowed to travel north by land back to New England. The 1873 voyage was his first since the Civil War kept the fleet in port, and Soper had his unexpected entry in the history books.

The four first-person accounts of 19th century whaling in this book—the never before published "The Voyage of the *F.H. Moore*" by Sam Williams, and excerpts from three works first published in the 1800s—open a world long closed to us by the passage of time. They reveal the human side of an inhumane trade, documenting both the best and worst of the whaling men and the times they lived in.

Prior to my discovering the manuscript of "The Voyage of the *F.H. Moore*" and the unfinished fragments of Williams' other writings, my knowledge of whaling was largely confined to watching the film *Moby Dick*. In reading and editing the story of the *Moore* I departed on my own voyage of discovery to times and places outside my experience. As a native Midwesterner, learning about matters nautical was like learning a foreign language, one in which I cannot claim fluency. But in researching the story of the *Moore*, a lost world opened to me, a world often ugly and repugnant, but also fascinating. Nineteenth-century whaling was an important chapter in American history. Much of it has been romanticized or distorted to cartoon-like dimensions, but it was for a time a part of the economic, cultural and political story of America projecting itself into the world.

Along with Williams' never before published account of his 1873–1874 voyage aboard the *Moore*, I was able to obtain a copy of the ship's logbook, now in the collection of Yale University. I gratefully acknowledge the help of Sam Lewbel, who photocopied the pages.

The following three excerpts come from previously published first-person accounts of 19th century whaling. The oldest of these is *Incidents of a Whaling Voyage* by Francis Allyn Olmsted, first published in 1841 by D. Appleton & Co. The next, *Etchings of a Whaling Cruise*, J. Ross Browne's expose and diatribe against the whaling industry after his hellish experience at sea, was first published by Harper & Brothers in 1846. The last of the excerpts is the most recent; written by Capt. Charles Henry Robbins, *The Gam: Being a Group of Whaling Stories* tells many stories of 19th century whaling, along with Robbins' personal story. First published by Newcomb & Goss in 1899 under the direction of his daughter Lizzie, a second edition was released in 1913. These works are in the public domain and are available in full online.

In editing these works, I have tried to leave certain words alone to reflect the usual way they were spelled in the 1800s. For example, people of that time spelled "coconut" as "cocoanut." I have also left alone the misspellings of place names because the fact that the writers used incorrect names says something about their attitudes toward the places and the people who lived there. Punctuation is also largely left alone. Most importantly, I have not sugarcoated the words or attitudes of the writers, which by modern standards are often racist or derogatory. Censoring the past to satisfy contemporary sensitivities is distorting history, and distorted history is not only useless, it is dangerous.

PART I: "THE VOYAGE OF THE *F.H. MOORE*"
BY SAMUEL GRANT WILLIAMS

Chapter One

Meet Fluker and the Crew

> *"I must go down to the seas again, to the lonely sea and the sky,*
> *And all I ask is a tall ship and a star to steer her by,*
> *And the wheel's kick and wind's song and the white sail shaking,*
> *And a gray mist on the sea's face and the gray dawn breaking."*
> —John Masefield, Sea Fever

After a lapse of forty-three years, I am reading over my old sea journal and find much that is interesting to me and perhaps it may interest others. On May 12th, 1873, the little brig *F.H. Moore*, captain Robert Soper, left Boston for a whaling voyage, to cruise among the West India Islands, and along the coast of British Honduras.

The *Moore* was owned in Boston by Sewell, Davis & Co., cordage dealers who had their place of business on Commercial Street in that city. Captain Soper was part owner of the *Moore* and it was said that his interest in the Brig was all he had left of a once comfortable fortune, acquired in the whaling business and which he had lost in the coal oil business in Pennsylvania. Captain Soper when he sailed on this voyage was about sixty-five years of age, a little old man, quite well preserved and fairly active, a very pleasant old man, a fine navigator and a good whaleman. He and his father before him had made successful whaling voyages in the Gulf of Mexico and about the West Indies, and it was to recoup his lost fortune that this voyage was planned and carried to a finish.

Our first mate's name was Elisha P. Huzzey; he was from Beverly, Massachusetts. He too was an old man, well over sixty but active as a boy of twenty-one and a pleasant, kind man. He was a good whaleman but a poor disciplinarian, and the crew often took advantage of that fact, but no evil effects ever resulted from his laxity, although an occasional outbreak jarred the otherwise good conduct of the crew.

Charley Tucker of New Bedford was the second mate. A big, good-natured fellow, not much of a whaleman, or at least he did not show up very well in a boat, but a thoroughly companionable man and a capital sailor. He "knew the ropes" as the sailors say, and kept the crew in their places. He left before the voyage was half-finished and the mate's son Henry Huzzey took his place. Young Huzzey was neither a sailor nor whaleman and was very little thought of by either the captain or crew.

The third mate was named Anson Streeter. He hailed from Vermont; he was the best whaleman aboard the Moore, but from the time the brig reached port till we were back again in blue water and the old third mate had exhausted his stock of booze, he was, to use a sailors' phrase, "three sheets in the wind and the fourth flapping." Although an awful old drunk when in port, he was companionable at sea and we all liked him. He caught the most of the whales on the voyage and was for that reason called lucky, but after years of experience I came to the conclusion that Streeter's good luck was brought about through carefully-planned work in the boat, and following his methods gave me many a whale I would not have gotten otherwise.

The boatsteerers were Horace Peas and Louis' dePieno, two Brava, Cape Verde Islands, Portuguese, and the writer, a down-east Yankee from Portsmouth, N.H. In 1870, most of the whaling vessels were supplied with crews by one S.J. Fluker, a Commercial Street, Boston, shipping agent. I never knew whether his right name was Fluker or not, but if it was it was an appropriate one.

How well I remember Fluker's little office. A flight of narrow, dirty stairs led one up to a couple of small rooms, one of which was reserved for sailors and greenhands, the other for the captains and the officers. The shipping agent had his desk and office with the Captains. If those four walls could only talk what stories they could repeat, for Fluker's office was a famous resort for the old captains and mates to gather, and many an hour was given up to sea yarns by these worthies. The sailors seldom got in there, unless it was to "sign articles," and then they hardly had time to look about before they were hustled out by Fluker.

Fluker was a "hot air" peddler of the first class. He knew the weakness of the average greenhand to a nicety and had a ready answer for every question. He knew the record of every Captain and mate in the whaling business, and according to Fluker they were the most fatherly and brotherly lot of fellows who sailed the seven seas. Fluker also had the interest of every boy at heart and, to use his own words, he "would as life, ship his own

One. Meet Fluker and the Crew

brother with any one of the skippers whose ships he filled with boys who he knew from experience would hardly return in the same vessel."

Of course the boys wanted to know how they lived on the Whalers. Fluker could tell them, "Boys the F.H. Moore is the best fitted out ship that sails from Boston. Captain Soper gives his men just what he eats himself, and just as plenty."

One of the greenhands wanted to know if they would eat pie on board the Moore.

"Pie, why fellows, you get your ration of pie every day. But boys, when you are down in the boats whaling, you will have to eat your mince pie cold."

The boys thought they could stand cold mince pie if it had plenty raisins in it. What a cruel disappointment they met with when they were clear of Boston Bay. No mince pie cold or hot. However, there was plenty of good wholesome food on board the Moore and the crew had little cause to complain. Captain Soper liked to eat and he frequently made port, and there was plenty of bananas, oranges, cocoanuts and pineapples, together with yams and sweet potatoes. While we were around the West Indies we frequently got large catches of fresh fish and turtle. So take it altogether we were well fed, minus the pie.

"I must go down to the seas again,
For the call of the running tide is a wild call and a clear call that may not be denied,
And all I ask is a windy day with the white clouds flying,
And the flung spray and the blown spume, and the seagulls crying."

In 1873 there was a little sailor's outfitting shop on Commercial Street in Boston where Fluker used to take the crews he shipped in the whalers and outfitted them for their trip at sea. The outfits usually consisted of a couple of cotton shirts, and a couple of woolen shirts, a pair of woolen pants, a couple of denim pants, a straw bed, a blanket, straw pillow, a pair of low shoes, pan, pot, spoon, knife and fork, a sheath knife and belt, a straw hat, two or three pounds of plug tobacco, two or three clay pipes, needles and a hank of thread, some soap and a towel.

This was supposed to be enough for a starter. All of the whalers carried "slop chests," that is, extra clothing and supplies, which were, from time to time, doled out to the crew, if they had anything coming to them from the oil taken. If luck was poor and no whales were taken, the sailors got very little from this source. However, the crews of such vessels, in time, got to looking like scarecrows.

How well I remember that little outfitting shop and the little old man who operated it. Its shelves filled with the cheapest sort of clothing, boots and shoes. Its festoons of all clothing, "Genuine Cape Cod Oil Clothing" guaranteed to shed water like a duck's back, but which in a short time absorbed the moisture like a sponge.

One of the boys wanted a suit, but it would make his bill too large; Fluker was looking out for that. He never forgot those little essentials. "Why my boy, you don't need any oil skins. You are going down in the tropics where it never rains, and if an occasional doldrum does strike you it's as warm as dish water." The little old man, who looked as if he had absorbed a large quantity of the mixture with which the oil skins were painted, assented. But it was plain to be seen, he did not have the owner's interest quite as much at heart as Fluker did, and would gladly have included an oil suit for everyone of our crew.

Many a wet, cold night the boys thought of that little store and its strings of oil clothes, and cursed Fluker loudly. They were learning the ropes, and that was one of the lessons, and not a pleasant one. For a wet back is not a pleasant sensation, even when you have plenty of warm clothes to change with when your watch goes below, which we did not have.

"I must go down to the seas again to this vagrant gypsy life,
To the gull's way and the whale's way where the wind's like a whetted knife
And all I ask is a merry yarn from a laughing, fellow-rover,
And a quiet sleep and sweet dream when the long trick's over"

Chapter Two

Our First Day at Sea: Goodbye to Boston

Dear Old Boston. It is hard to leave the old "Hub" anytime, but we could not have picked a more sublime day for our departure than it was on this May morning, when we hove our anchor up out of the muddy bottom of the bay and with our pilot walking the breakdeck set all sail and, with a gentle breeze from the west, worked our way down the bay.

About noon, we bid goodbye to the pilot and some of the owners who left us to board a tug, which was to take them back to the city. Bracing forward the yards, we were off for the broad Atlantic. The noon meal, the first to be eaten on the Moore, was consumed in silence by us younger men, but the Captain and mates talked about whales and whaling grounds while us youngsters listened.

After dinner we watched Boston fading away in the distance. One by one the objects so dear to us faded away, the gilded dome of the State House being one of the last.

In the afternoon, we unbent the cables from the anchors and stowed the cable in its lockers and lashed the anchors to the rails. Massachusetts Bay was alive with shipping of all kinds; steamers, schooners, fishermen and yachts. Some of the yacht clubs were trying out their new boats with their old ones, and they presented a fine marine picture. Such tremendous sails and such shining white hulls and finely kept rigging. Their crews in their natty white suits, and the owners and their guests in yacht-costumes. Several of the yachts passed quite near us, and I guess we were as much of an object of curiosity to them as they were to us.

Talk about being seasick, it seemed as if the whole crew had the malde-mere when we struck rough water and the breeze freshened. Even Captain Soper was sick and made frequent trips to the side to unload. Those who were not sick had the blues and all wanted to get out of sight of the land and have it over.

Part I—"The Voyage of the *F.H. Moore*"

Clearing Cape Cod, our course lay South-South-East, and after choosing watches the Port watch went below at eight o'clock.

"Watches?" we hear you say. Yes, the whole ship's crew is divided into two parties, and the party headed by the Captain is called the starboard watch. The second mate has charge of this watch; the other is called the port watch and is the first mate's gang. The Starboard watch always takes the first four hours on deck when leaving port, after that, they take turns about automatically. I remember Highland light and the lights off the *Nausett*.

They shown out bright all through our watch, but when we returned to the deck at four o'clock the next morning, there was no lights or land in sight, nothing but green water. We did not strike blue water until well into the afternoon next day. Captain Soper was in a hurry to get off shore and onto whaling ground, and we carried sail pretty hard, and that made it disagreeable for the seasick crew. But after a few days we got over that, or at least most all did. Two or three of the boys had a bad time of it however, taking several weeks to get their stomachs settled.

There was plenty to do, so much for the boys to learn. They wondered if they would ever learn all the ropes the compass and sea terms. Learning to walk the deck of a little vessel of 106 tons is no easy matter to the greenhand, for a vessel of that size pitches amid rolls much quicker than a large ship, but it is these little ships that make good sailors.

During the day the breeze calmed down and left us later with our sails flapping against the masts. We were now getting out into the Gulf Stream and were drifting in a Northeasterly direction with its current at a rate of two miles an hour. It was no use for Captain Soper to walk the deck and whistle for wind, and for once the captain had to wait, and many times after I saw that impatient look on his face.

"But what was the use of worrying over a calm the second day out when you had a year and a half of weather, good, bad and indifferent ahead of you." said the old third mate.

Becalmed in the Gulf Stream. There we were drifting towards the banks of New Foundland. Our course in an almost opposite direction and our voyage before us, and a long one at that. Think of eighteen months to be spent in a little craft, 106 feet long and 22 feet wide, and then wonder if you will, that friend-ships lasting through life are made, and that here too, bitter quarrels find a breeding place, which end only in the separation of the interested parties.

Two. Our First Day at Sea

Yet go to New Bedford and interview the old "Nimrod of the Sea," and talk to him of an eighteen months voyage and he will sniff and call it a "plum pudding voyage," or "only a dog's watch." The dog-watch is the short watch in the evening between four and eight o'clock, which is divided into two watches to bring about the change in watches, so that one watch will not be on deck eight hours every night. Hence the old sailor who was used to going for three and four year voyages did not think our voyage worthwhile. But eighteen months is some voyage after all and most of us thought so.

Any boy who thinks there is nothing to do on board of a ship at sea once she is clear of port with her cargo under hatches, is sure to meet with a surprise, and especially if he casts his lot with a whaleman, for not only are there all the ship's duties to be performed, but there are three or four boats to be fitted up for the chase and that means a lot of extra work, for nothing must be left undone.

When the whales are sighted, everything must be in readiness. Every preparation must be made for the supreme business of the voyage. Every boat has five pulling oars and a lone steering oar, the latter for use when the weather is calm when it would not be possible to pull on a whale, as the splash of an oar would disturb him, for whales can hear or sense danger a long distance. His hearing or instinct of danger is very acute.

There are two big tubs into which the line is coiled—some 300 fathoms, or 1800 feet. This line is about ¾ of an inch in diameter and is made from the finest of imported manila. It is soft as a rag, but strong as wire. Each boat has two fresh water kegs, which hold about five gallons each.

Then we always had a big bag of bread—no mince pies—and that was the limit of our "grub-kit." Five irons or harpoons and three lances, a bomb gun and half a dozen bomb-lances, two flags or waifs to use when the whale was dead to call the ship. A lantern, tobacco and pipes in a water-tight keg, in case we were caught out after dark, a hatchet, a big boat sail, two knives, and a small boat spade, and many other little things that come handy.

All this gear had to be kept in first-class order and always ready for instant use. We could clear and drop our boats into the water in a very few minutes, and after our crews had been six months at sea they got to be expert in getting in and out of a boat. So you see there was no loafing on board the Moore from the first day out. Captain Soper believed in giving his men "plenty to eat and plenty to do." He argued that busy people find small time to growl and find fault.

Part I—"The Voyage of the *F.H. Moore*"

All that day we worked on our boats and gear and about four p. m., a light breeze sprang up from the West and our little brig, with all sail set, once more moved off towards the Southland. One of the crew, sent aloft to loose the top-gallant-sail, sent a thrill through the crew by cry of "sail ho," and shortly we saw her from the deck, a handsome full-rigged ship, bound from New York to Europe. She had all sail set and was booming along and we passed without signaling, so intent were we in getting along about our business.

Nothing of further importance happened that day, unless it was the fact that the crew were getting over their sea sickness, and getting broke in to their work. It was quite a while before all the greenhands could go aloft, or take their "trick at the wheel" but in time they all could steer some, although some of the boys ware a long time learning the compass, which one finds is much "harder to tell time by than the watch," as one of the boys remarked.

Editor's Note

The handwritten logbook of the *F.H. Moore* begins, "This is a journal of the Good Brig F. H. Moore which sailed from Boston May 12th, 1873 on a Whaling Voyage under Command of Captain. Robert Soper. This day Began with moderate winds from the N.W. and Fine weather. Took our Anchor at [illegible.] Chose watches and commenced getting the Boats ready..."

On most days, the entries in the logbook were short and routine, almost always starting with a record of the weather and wind. Capitalization was random and punctuation was often nonexistent. The text was illustrated with drawings of whales spotted by the crew with notations of the number of barrels of oil they yielded. On Friday the 16th of May, the logbook, in part, read "raised Sperm whale and lowered but did not strike shortened sail this night and stood quarter watches Day light called all hands and had all sail."

Later in May, the logbook read, in part, "Raised Sperm whales got Breakfast and lowered all boats and chased until 3 p.m. Starboard boat struck twice and lost them so ends this day."

"So ends this day" was a frequent entry in the logbook during the long voyage.

Chapter Three

On the Way to San Domingo

In my last chapter when I left you, we had caught a good breeze and were moving to the south. We had a fair wind and fine weather for two weeks and we are now in the Bahama Islands, the extensive group of small islands and keys lying Southeast of Florida.

One of these groups of islands, now called Watling Island, was the first land discovered by Columbus in 1492. It was among this group of islands and sand keys that the pirates made their home in 1600, and one of the islands, Harbor Island by name, was settled by the descendants of the buccaneers. These islands afforded a shelter for the blockade-runners at the time of the Civil War and the business of Nassau, the capital of the group, grew from about $1,000,000 to about $26,000,000 in a very short time. At the end of the war, however, business fell off again and at the present time it consists principally in supplying Southern states with such fruits as oranges, pineapples and cocoanuts. The winters are fine and many American tourists winter here. But from early in September till November there are hurricanes and heavy gales that sweep over the islands and render the Bahamas undesirable.

We had a fine, fair wind from the Northwest and did not land at any of the Bahamas. We ran down through what is known as the Caicos Channel. These grounds to tarry in any length of time must have been dangerous in the early days when they were uncharted. They are full of little reefs and dangerous currents, and are not much sought by the whalers. So we left the land of Captain Kidd and his companions for the pleasanter whaling grounds further south.

Our passage from Nassau to Porto Prince bay in the island of San Domingo, down through the Caicos passage, or channel, as it is sometimes called, was made with a strong northeast wind blowing. Here we made our first acquaintance of a water-spout, and as we passed quite near one I took particularly care to watch it as did Captain Soper, for they are not pleasant neighbors and have been known to dismantle a ship. This one was some two miles distant I should judge and extended from a black, threatening cloud to the water, a long, narrow white fillimish looking monster, if I might call it such. The sea at its base seemed to be boiling, reminding me of a great whale breeching, the water flying in every direction. Now and again the lone spout would weave and oscillate and a dark ring would travel up and down it. At last the spout became fainter as the great rain cloud discharged its torrent of water into the ocean. We were treated to the sight of a number of them before we got out of this region and one can hardly appreciate them till they find themselves in their neighborhood.

At eleven in the morning we sighted through the rain squall at a long distance the large island of Mayaguano. It was too far off to be seen very plain and looked gray and forbidding. Our course steered through the passage was south-by-east, and as our brig cut through the emerald green water, dotted with an occasional sail, the distant gray land, the bright, silver colored bonita, a beautiful fish gliding through the water by our ship's side, and voraciously jumping at the rag baited hooks that the sailors threw to them, and the heavy clouds scudding past us, presented a picture not soon forgotten. We caught all the bonita we wanted and besides eating all we wanted fresh, salted down a lot. The Caicos passage is one of the finest places to tryout a ship's sailing qualities that we know of. The wind is steady and you can carry sail as hard as you want to. At two p.m. we sighted West Caicos and passing by that island, which was a port of refuge for some years, we worked down towards the larger island of Inagua, and at sundown West Caicos bore by the compass east-north-east and Inagua south-south-west. The latitude today was 21 deg. 27 min. north, longitude 73 deg. 37 min. west from Greenwich.

Early next morning we made the island of Tortuga and passing to the west of it made San Domingo, passing down the western end and then turning east we soon came to Porto Prince Bay, at the head of which is quite an enterprising city of the same name under the rule of the Hytian Government, the greatest place on earth for rebellions and a land of sugar and molasses.

The bay of Porto Prince is filled with sugar drogers, little vessels loaded with sugar bound for America and Europe. Their white sails lend a charm to this beautiful sheet of water. The native fishing boats are also plentiful, and as we pass near by one we notice the crew are perfectly nude. They do not wear clothes when fishing but put them on when they return to port.

Three. On the Way to San Domingo

These boats are quite large and carry immense latine sails, all out of proportion to the size of their boats, but they are good sailors are these Haitian fishermen and carry sail hard. Some of these boats are loaded with flying fish, the daintiest fish in the sea. They are said to subsist on seafoam.

The mode of fishing for them is to set their sail at night and drift with two lights set well above the hull of the boat. The flying fish fly for the light, strike the sail, and fall into the boat. Coming into the bay from the sea one is struck by the beautiful green hills of the island. The coast of Ireland is the only place we ever visited that has this beautiful emerald setting. Once seen it is never forgotten. You find it nowhere else on earth it is said. Certainly Ireland has a close competitor for beauty in San Domingo.

High up in the mountains one sees little columns of white smoke. It belongs to some sugar plantation or rum still, and one can almost fancy they smell the rum, which is made in large quantities here, when the breeze is blowing off shore. The waters were alive with all kinds of fish and great flocks of birds were everywhere. Porpoises and cow fish, the latter a species of porpoise, and the whalers would rather have a porpoise anytime than pig. There were great flights of pelicans, the awkwardest bird that flies. It is fun watching the pelicans fish. We he sees a fish he lets go and comes down into the water like a thousand bricks. Then you see them rising slowly out of the water, the great bag under their bill hangs like a cow's udder.

They are strong fliers, however, and they make long journeys from the land in search of food. Occasionally the Frigate bird gets after a pelican and sometimes will make the pelican disgorge his bag of fish. Then the Frigate bird will catch the fish before it strikes the water. That is if the fight is waged any great distance in the air. Sometimes as many as one hundred pelicans are in sight at one time, and when they are all fishing they keep the water pretty well churned. The little land birds one sees here are of a beautiful plumage, but they do not sing like birds in the north.

We took in sail when we reached Porto Prince Bay and are to cruise here for a time as Captain Soper says there were whales here in years gone by.

We have been cruising in Porto Prince Bay for two weeks now. It is a big sheet of water, some thirty miles across, and sperm whales are sometimes taken in the bay close to the anchorage. This is one of the loveliest places on earth. It reminds me of a verse in the Missionary Hymn, where it says of India: "Where every prospect pleases and only man is vile." We are now standing boat-crew watches. That is, the night is divided into three watches of four hours each and each of the boat crews has four hours out. We take in sail every night and we all get plenty of sleep, in fact more than we need. But we are all on deck early in the morning and spend the day at ships duty. The crews have been drilling with the boats and every fine day we lower all three of the boats in the evening and take a good hard pull to harden up the muscles of the crew and give them plenty of practice. Our boys are getting so they can pull for an hour at a time, but they growl now if the pulling is continued too long. The mate has the best boat crew for pulling. He has proved that on several evenings by beating the second third mates crews. But there is not such a great difference in crew for the first mile. But we are all glad to get back to the brig and rest after a two mile pull. We have men early at the mastheads early in the morning and there are five of us on the lookout for whales all the day long. Nothing escapes us.

Captain Soper has offered a $5.00 gold piece as a gift to the first man to raise whales. So we are all alert. Evenings the boys in the forecastle sing and dance and they have a fiddle. They seem to be enjoying themselves. Us boatsteerers and the mates have our good times. We tell stories, read a good deal, do some studying and Captain Soper, who is a fine navigator, gives us lessons in navigation. I could find the latitude at noon by the sun, but Captain Soper has taught me to find the latitude by the North Star. Some of the stories of adventure told by the mates are worth retelling. Mr. Hussy was sitting on the after hatch the other evening and told this story about a former shipmate of his.

Chapter Four

Story of Captain Peakes

Not long ago I was looking over an old Boston newspaper. Naturally I turned to the shipping news and the first thing I noticed was the following:

"LOST off Marthy's Vineyard in the gale of Tuesday, the little whaling bark *Sarah*, of New Bedford, Captain Peaks. The *Sarah* was two days out from New Bedford and was capsized during the gale. Captain Peaks went down with his ship. Three men belonging to the crew were picked up the next day from the wreckage by a pilot boat and brought into New Bedford."

"This item brings back to me." said Mr. Huzzey, a flood of recollections of the dead Captain and his little old fashioned bark. I had known him for years. He was considered one of the luckiest whalemen sailing from New Bedford, but he must have sailed on a Friday on this voyage in the *Sarah*.

Whaleman who sailed from New Bedford in 1870 remembered him as a man of about fifty, broad shouldered and active, an aggressive fellow, with blue eyes and a big head covered with a thatch of coarse red hair. In fact some of the skippers used to call him "Red Peaks." He had one little bald spot, about the size of a silver dollar, which gave him a rather queer look, and was often the cause of a joke among his friends. He explained it away by saying: "I got the hair worn off in that spot when I was a boy on my first voyage, hunting around in my sea chest for my clothing.' In explanation I will say that greenhands sailing from New Bedford often complained of the scarcity of the clothing received with their outfit when they sailed, and it got to be a saying among them that what one didn't find in his chest, he could always find in his bill." Captain Peaks was a capital storyteller, and I often listened to him. I remember one of his stories, it was something like this:

Part I—"The Voyage of the *F.H. Moore*"

How Peakes Caught the Whale

"In 1869 or thereabouts, I was second mate of the little bark Edward Everett, of New Bedford, Jake Black, of Edgarton, was master, and Bob Macy of Nantucket, was first mate. We were cruising off Juan Fernandez on the coast of Chile. It is a very rough place to cruise in the months of July and August, and all the whalers but us had left that ground for fairer weather on the Off Shore Grounds, some 500 miles west of the coast of Peru. We had taken 300 barrels of sperm oil since leaving Conception Bay and Captain Black was anxious to get another big "squarehead" before leaving the coast of Chile. But the weather was bad, I never saw any such a continuance of bad weather and the "old Man" had decided that if there was no improvement in another week he would square yards for the "Off Shore Grounds.

"It had been blowing very hard all the night before, and when the morning broke it found us laying under very short sail, with our decks wet with a constant shower of spray that was coming over the weather rail and the crew of the watch on deck all sitting under the after house trying to keep dry. Off to windward lay Juan Fernandez some twenty miles away and between us and the island there rolled as ugly a sea as you ever saw.

"I had just ordered the men into the rigging to see if any sails were in sight, and the boatsteerer had no more than got his feet on the shoerpole when he shouted to me, 'Mr. Peaks, there's a hundred barrel sperm whale right under the vessel's stern.' I ran to the tafrail and my eyes almost popped out of my head, for there, not two hundred yards astern was as handsome an old squarehead as you ever saw. He was heading to leaward and the waves rolled over his long, brown body and never a one broke. He had gone through our wake and did not notice it.

"At any other time a whale would have been gallied and would have shown us his flukes. I jumped to the cabin companionway and shouted to the Captain that there was a big whale under the stern. He awoke from a sound sleep, and not knowing the state of the weather, shouted to me to 'hoist and swing' and see if I could not strike him before he went down. I didn't stop to think about the weather, either, but just called my crew and the next minute we were lowering the boat.

"How we ever got clear of the ship is more than I can tell, but we did, and the next thing we knew we were clear from the ship astern and with oars out were pulling after the whale. My boat a 30-foot cedar, one of

Four. Story of Captain Peakes

Leonard's make, with high gunwales and as light a cork, rose on every wave and made splendid weather of it.

"We soon came up with the big fellow, the crew saw that we were all right, and before I could realize it I had pulled up alongside of him and my boatsteerer had darted two irons into him. There was a splash of his big flukes and he was gone and the line was flying out of the tub.

"My boat swung around and took in a lot of water in doing so but the crew grabbed the boat buckets and bailed for dear life. Then I looked at the ship for the first time since I had lowered. The Captain was standing on the tafrail swinging his hat. He afterwards told me he had been shouting to me to come back but I thought he was encouraging me to hold on and I expected to see Bob Macy lower his boat every minute, but he didn't and Captain Black afterwards told me 'He wasn't going to lose but one fool' and had hard work keeping Bob from lowering too.

Fortunately the whale did not run, as some whales do but he lay and tried to roll the irons out of him. This gave me the finest chance in the world to kill him and I soon had him spouting thick blood. It was an anxious period from that time until he died. If he run I know I must lose him, but luck was with me again, and he died under water. Very soon he floated to the surface and I found I could keep in smooth water by keeping to leaward of him.

Now that my whale was dead the next question was to save him. It was a sure thing that I could not lay by him any length of time, and to tell the truth I wanted to get on board the bark the most of any of my boats crew, for I both knew and felt the danger we were in. I knew that Captain Black would come to me or carry the masts out of the Edward Everett. He was working up to windward of us and we could see the seas breaking over the barks bow as she pounded into them.

I knew we had a new two inch hawser on board and if we could only get that fastened to Mr. Whale's tail, we stood a chance to save him. I also knew that no fluke chain ever made would hold him. But to get a two inch hawser fastened to a whale's flukes in a gale of wind was the question.

"Captain Black had no idea that he could save the whale, his only thought was to save a boatload of "fools" as he always called us, but Bob Macy had some hopes, and had against the Captain's remonstrance, brought the big coil of rope on deck and making one end fast to the foremast, got ready for the boat to come alongside.

"It took a long time for the old bark to get to me, at least it seemed so, but I knew the Captain was driving her for all she was worth. At last I saw the bark wearing ship, she could not tack in such weather. As she came to the wind heading towards me I saw her ship a big sea. I held my breath. Could she right herself under the weight of water on her deck? It was a happy moment when I saw her right and punch her old blunt nose into another sea.

"On she came and in a few minutes I was running my line to the bark. Just then the bark's main topsail gave a slap and the whole sail ripped completely out of the boltropes, and hung in the lea rigging. It was a moment of excitement. Captain Black shouted 'Never mind the whale, grab these lines and save yourselves.' And Bob Macy shouts 'Try and get this line about his flukes Mr. Peaks.'

"I heard Bob and threw my line to Macy. I grabbed the end of the hawser and my crew commenced hauling back to the whale. It took a little time to get the hawser around the whales flukes, but I did do it, although I never can tell how it was done, and I got a good hitch on it too, and just then I found myself in the water and as I could not swim very well I grabbed the first thing I could get hold of. It was the line I had run to the ship and I commenced to haul myself along it towards the bark.

"I could see my crew struggling in the water, but fortunately they all were good swimmers and reached the bark before I did, where they were hauled aboard by their shipmates. I was the last to get on board and was dipped up and down several times by the line coming taught and then slacking up. But I finally got near enough to the bark to catch a bowline which was thrown to me. There was a half a dozen of them, thrown. Some went past me, but I grabbed the nearest and went out of sight under water as I let go of the hauling line. But I felt my lifeline tighten and the boys in their haste pretty nearly pulled it out of my hands, but the next moment I got the bowline over my [word missing from the text] and shoulders and before you could say "hoist away" I was being hauled over the rail and Bob Macy was pounding me on the back and the Captain was jumping up and down and shouting in my ear 'Did you get a good knot on that hawser?'

"My handsome boat was floating away off to windward with everything belonging to her. When the line came taught under it capsized her and that accounted for our being in the swim. We lost her and never saw any signs of her, although we spent several days later on trying to find her.

"But we saved the whale. In fact after we had fastened him to the ship he acted as a sea anchor and we rode quite nicely to leaward of him, the hawser held, and a few hours later it commenced to moderate. The next day it was pleasant enough to cut him in, and later when we boiled him out he made us over 100 barrels of sperm oil. That was the first and only time I ever lowered for a whale in a gale. I have seen sperm whales several times since when it was too rough to lower for them, but I have had my lesson. I don't want any whales under any such circumstances."

Chapter Five

Gonives Bay and Our First Whale

At the Eastern end of Porto Prince Bay is a finely sheltered but smaller bay known as Gonives Bay. Then there is Gonives Island, a small island the home of a few poor fishermen. We cruised near the island and it was here we captured our first whale. I remember the circumstances well. The day was a fine one and we had just finished our noon meal, when the cry "there she blows," rang out from aloft and electrified the whole ship's crew.

Talk about excitement, everybody was on deck in short order. The mates climbed aloft with their glasses and it was soon decided that the whales were a school of sperm whales, and that they were working off shore before the wind, which gave us a chance to get in behind them. The brig was brought up into the wind, the foresail was brailed up and the yards hauled aback, to stop her headway. Then the order was given to get the boats ready. But we did not lower the boat's right away as we wanted to see how the whales were acting, to get their run. So we watched them for almost an hour and then the boats were carefully lowered. The mate, second mate and third mate immediately set their boat sails and headed in the direction of the whales, the brig signaling their position.

We did not see them from our boats during that "rising" but each boatheader took a course which he thought would bring him as near to the submerged whales as it was safe to go and not run over them, for to do so would be to frighten or "gally" them. This laying for whales is a fine art in a boatheader. His success as a whaleman is largely due to his knowledge of whales and their habits. To be able to forecast where the whale under water is going to reappear, and to get within striking distance of them and not frighten them. Of course he must know how to kill his whale once he is fast to one, but finding them and getting fast first, and in that way securing the most whales for his boat, is what spells success for a whaleman.

Five. Gonives Bay and Our First Whale

Now Mr. Streeter was as I have stated before skillful in this kind of work. He had watched this school of whales from the time they were raised and he proved that he had got their course a little more accurately than the other two mates, for he worked his boat away from the other boats fully half a mile and when the brig signaled that the whales were up, Mr. Streeter told me to stand up on the bow of our boat and see if I could find them.

I had hardly mounted on the bow before I saw them. Mr. Streeter saw them too from his lookout on the stern of the boat. He jumped down, hauled aft his sheet and telling the crew to take their paddles, swung the boats head off in the direction of our prey. The whales were not going very much, in fact they were making a circle when they first came up as sperm whales often do when they first came to the surface. We came down to them fast and as they were lying "heads and points" we slacked off our sheets again to give them a chance to straighten out and start off on a course.

This they finally did and as soon as they were once more moving we again headed our boat for them. They did not go very fast and we overhauled them rapidly and steered for the hind one in the bunch, a big bull. He was lagging behind the rest fully fifty yards and we ran in behind him. He was a handsome, big brute and we could hear him spout some distance. There was a nice breeze and we overhauled him fast. I kept my eyes on Mr. Streeter and soon I got the signal from him to lay down my paddle and get ready to strike the whale, that is dart the iron into him. As I gained my feet in the bow I grabbed my iron, shore people call it the harpoon, and waited for Mr. Streeter to say "when."

It was an intense moment. I could see the big fellow's flukes under the water. I knew if he hit the boat with that tail of his we would have a stove boat and might have to swim. An instant more and he comes up to spout, our boat shoots up broadside to him and not ten feet distant. "Give it to him, Sam" whispered Streeter, and I slammed the iron into him. I saw the iron bury itself over two feet in his brown side and grasped the second iron, "give him another Sam" yelled Streeter, and I did, the second iron burying itself well forward of his hump. There was sudden submergence of the great hulk, a splash of water from his flukes and he disappeared, I throw the boxwarp overboard and turned around to hear Streeter holler, "Roll up that sail."

The line was flying around the loggerhead, and burning into the solid oak post used for snubbing and the after oarsman was throwing water into the line tub to wet the line. The boat was going through the water fully five or six miles an hour and Streeter was hanging onto every inch of line, for he knew it would all have to be hauled in again. But our whale did not run far and soon came up on top of the water and commenced to roll. The boat sail was taken down and stowed away, and I changed places with Mr. Streeter, whose business was to kill the whale, while mine was to steer to boat and see that the crew pulled the boat up close enough for him to do so.

Our whale was doing his best to roll those irons out of his hide, and the contortions that he went through was wonderful. He was the most animated piece of flesh and blood in sight and to the greenhands he was a terrifying spectacle. But to an old whaleman like Streeter, who had killed a hundred or more in his lifetime he was nothing but a whale and a big clumsy lout of a fellow at that. We haul line to the whale and when within about fifty feet of the big lump of blubber, we take our oars and pull for him. Streeter has his bombgun ready. "Put me a little further forward, Sam" and I swing the boat towards to whales head,

Streeter waits patiently. It seemed that he never would fire. Presently the whale rolls just right, the third mate brings the big gun quickly to his shoulder and fires, the recoil of the big gun knocks him sprawling in the bottom of the boat, and I hear him yelling to "stern all," and we are backing away. There is a big shivering of the great hulk, a muffled explosion is heard, it is the bomb exploding in the whale's lung and he settles down out of sight and the line is once more flying around the logger-head. "Hold on to the line, Sam " councils Streeter, and I snub it so hard that I bring the boat's head down to the water, and the old mate yells to me not to "swamp" her.

Our whale stays a long time under water. The other mates come pulling up to our boat and we all wait. "When he comes up he will be a dead whale" says Mr. Streeter. "I got a good chance to put a bomb into his lungs. "At last the line begins to go out slower and slower. The whale is coming up but the line is bowing under water. At last we see him. He is a mighty sick whale. There is no more fight in him. The thick clots of blood are rolling out of his spouthole. He is slowly smothering to death. His lungs have been torn to shreds by the bomb explosion. He does not know what he is doing now and soon he makes a long circle and rolls over on his side, his big jaw, studded with powerful teeth, drops down, and he is dead.

He made us sixty barrels of oil.

Five. Gonives Bay and Our First Whale

We had drifted fully forty miles to leaward from where we caught our whale before we had him cut in on the next day, and it took us twenty-four hours to work back to Gonives and then we anchored in a good safe place and boiled out our whale and stowed the oil away in the big casks in the hold. Then we cleaned the decks and all the paint work of the brig. We took the ashes from under the furnace and made a strong lye for this purpose, and by Sunday the brig looked as clean as a pilot boat and was ready for another whale. Captain Soper allowed no unnecessary work on that Lord's Day, but we stood "mastheads" when we were cruising and if we saw whales on Sunday we lowered for them, but we took no whales that I remember on that day.

But this particular Sunday Captain Soper allowed us to take a run ashore and get what enjoyment we could out of Gonives. Now Gonives island was at that time about as near "Poverty Flats" as it is possible for a community to be and its people exist. The inhabitants were all black and nearly nude and fish and yams, rice and a little fruit from the mainland of Haiti was all they had to live on. Children under ten wore no clothing whatever and their elders were little better clothed. I saw one old woman dressed in gunny cloth, and whenever I read about the witches in Macbeth I think of the old black crow of Gonives and fancy I saw a West Indian witch. By the way, the West Indian Negroes are strong believers in witchcraft and they tell hair-raising stories about the capers of the "jumbies."

The homes of the fishermen of Gonives were mere shelters from the sun and rain. It was a wonder that their owners could keep comfortable in such shacks. The old saying about a fisherman's luck must have originated here. They seemed contented, however, and I saw one young black fellow who stood fully six feet in his enormous feet, and who only needed good feeding for a few months to have developed into a second Jack Johnson. He sang, laughed and danced and appeared happy as if he had everything in the world and nothing to worry about.

There is an old saying that the poorer the family the more dogs it possesses, and these Gonives fishermen had dogs too. Hungry, wolfish looking animals who lived on what scraps they could get. They were as poor as their owners in every sense of the word. We made arrangements with the men to bring us fish and exchanged shipbread and meat with them. They certainly had a change of diet while we lay at anchor. There was a low, stunted growth of ironwood trees on the island and we got a lot of poles for our harpoons. We scraped the brigs' bottom before we hoisted anchor, and when we left all the inhabitants of Gonives Island including the witch- were down to the beach to see us off. Going out we passed an uncharted reef. There was about two fathoms of water on it. The fishermen had warned us about it, however, and we gave it a wide berth.

Chapter Six

Off for More Whales

Wednesday morning, one day out from Gonives. Fair weather and a light breeze from up the bay, which reminds me that we had been having a lot of the finest of weather, and that reminds me further of the old whaleman who had been enjoying a long spell of fine days and had made the following entry in his journal: "Fair weather and fine breeze" for a fortnight, when he concluded there was more or less of repetition, so the next entry read: "Weather—a continuation of the same."

The weather was fine and no mistake, and as we sailed to the southeast through the emerald waters of the bay, with our keen-eyed lookouts at masthead informing us of everything that was going on within their range of vision, very few of us cared leave the deck in the day time to sleep, for above decks everything was full of life and it was cool and comfortable from the trade winds, but below decks the air was hot and the bilgewater stunk us out.

The ocean seemed alive with fish of all kinds and to vary the monotony a large school of blackfish came in sight. But we did not lower the boats for them as we did not need the practice, but as they are a sign of Sperm whales sometimes, we followed them. A little later in the morning we sighted the town of Jermiee at the eastern end of Haiti. It is a small town which boasts of a good harbor, and coming out of it we saw a tall, handsome three-masted schooner.

We knew her to be an American by her general appearance and surmised that she might be bound for home, a fact in which we were right, she being off for Baltimore with a cargo of molasses. Having hailed her and put aboard letters for loved ones at home, we hoisted our boat and stood down along the coast. We cruised along until past Jerimee and not seeing any whales concluded to work down towards Jamaica and the Swan Islands.

From Jerimee to Jamaica our course was in a south-westerly direction and the passage was a pleasant one, with a leading breeze all the way. It

Part I—"The Voyage of the *F.H. Moore*"

took us two days to make the passage as we were on a good whaling ground and did not want to skim over the ground too fast.

Towards evening of the second day we came to anchor in South Negril, one of the lesser ports of Jamaica. The harbor was very small and rocky, on all sides great coral reefs rose above the water, grass clad and green with sea verdure. The tides of the West Indies are small, about two feet being the maximum, and Negril seemed to have no tide whatever. The people came off with their canoes loaded with fruit, and we traded pork, beef and flour and shipbread for yams, sweet potatoes, oranges, pineapples and coconuts. Of the latter, the trees surrounded the beach and harbor and were very numerous. Bananas were here in the heaviest of bunches; and all kinds of tropical fruit was abundant.

Some of the houses here were made of Mahogany logs a foot thick. If they could have been delivered in the United States they would have brought a great price. Just think of mahogany houses with. thatch roofs, and what with coconut trees constantly thrashing the roof to pieces and an army of rats building in them, one wonder was how they ever could have been kept rain proof, I guess they leaked.

The inhabitants were mostly of the pronounced black order. There were very few molasses colored niggers among them. A few poor whites, who looked as if they were all suffering from the yellow jaundice, were sandwiched in amongst the blacks. It was hard telling which was the laziest the blacks or the whites, they were all tarred with the same brush when I speak about their industry.

It was at this place that I saw some of the finest specimens of leaf coral. I saw one of a delicate pink, so beautiful, that it almost compared with a spiders web in texture. This specimen was owned by a wealthy sugar-planter and was a very rare and valuable specimen. Towards evening the black threatening clouds of a heavy trade wind gale began flying and Captain Soper not liking his anchorage hove up and we made sail and got out of Negril. We shortened sail after leaving the harbor and made good weather of the gale that night.

The Swan Islands

About 250 miles west and 60 miles south of South Negril are the Swan Islands. These little islands are little known to the American people,

Six. Off for More Whales

but they are owned by the New York Guano Company, or were at the time I visited them in 1874. There was a certain phosphate formation there, which was rich in fertilizing qualities, but the company owned other beds which were better than those of Swan Islands and consequently were not working these at that time. However, two men, Americans, were left in charge of them and here these two Crusos lived, monarchs of all they surveyed, but awfully lonesome and welcoming anyone who dropped anchor in the little boat harbor. We spent two days on the main island and enjoyed a most pleasant visit with Mr. Reed and his companion, whose name I have forgotten. There is a long strip of coarse white sand here which is the natural hatching place for the sea turtles in this neighborhood and at certain seasons of the year when the turtles come here to lay their eggs in the sand, the beach is a sight.

Mr. Reed told me of seeing hundreds of these turtles on the beach at one time. I did not get a taste of the Lenten luxuries at this time, but later on I was treated to a taste of them boiled and must say that I do not hanker for any more turtle eggs. They are strong and have a fishy taste that I did not like. They are good eating for those who like them but I can get along without them.

The company had years before placed a large drove of hogs on this island, which in time got to be expert in hunting turtle nests and eating the eggs. We were glad that they did have the eggs to fall back on for they would have starved without them.

Mr. Reed told of a case that came under his observation of a large hog attacking and killing a turtle which weighed 150 pounds. The first day we were there we were told if we could catch one of the wild hogs we could have him. Now a pig to a sailor is as good as chicken, better in fact as there is more flesh meat in a pig than chicken, consequently more good eating, so the crew of the Moore immediately started on a grand boar hunt. The sun was out in his brightest array, I might say it was boiling hot and not prevaricate, and after two hours spent in hunting among the rocky ledges of the Island and along the beaches without success we returned to the brig, thoroughly cooked with the heat and without having seen a hog. In the evening, however, Mr. Reed gave us one which he had in a pen. He proved to be the most savage specimen of the pork species we had ever seen.

We had him hog-tied when we brought him on board the brig, but when we turned him loose, expecting he would hide himself in the forecastle, he turned on us and drove us all into the rigging, and after we had spent an hour or more trying to pen him up and he had torn the pants off of several of the sailors who got too near him, we had to shoot him. We built a fire under the try-pots and soon had water to scald him and had fresh pork for dinner next day.

It was there that I saw a good joke played on "Goliah" one of our crew. Goliah was a big, black fellow and was a dead game sport. Mr. Reed was in the habit of filling a large tank with water daily for the hogs that came down in the night to drink, and who would after they had satisfied their thirst return to their hiding places. One large hog had wandered down to the trough in the day time, no doubt driven there by thirst and finding no water there had laid down by the trough and died. Goliah was called and promised this pig, which Mr. Reed told him was sleeping, if he could catch him. Supplying himself with a piece of rope to tie him with Goliah crept silently up behind the hog, stooping cautiously down he seized the dead pig by the hind legs, the joke hadn't even then penetrated his wooly pate, and he exclaimed, "Golly sah, dat pig is dead, Sah."

Mr. Reed gave Goliah a smaller pig for himself and the other boys in the forecastle and they had all the pork they wanted. Just think of living a year or more without hearing from home or in fact from any part of the civilized world. Wild hogs, sea birds and turtles as your companions. Fifty dollars a month room and board. Mr. Reed and his companion had a plenty to eat, plenty to wear, and they had lots of good books to read, but they were lonely. So when we bid them good bye they felt like parting with old friends.

We promised to come back again to see them but never was able to keep our promise. We never heard of them again.

Chapter Seven

Off This Coast of Honduras

The run from the Swan Islands to the Central American coast of British Honduras is a short one when your sails are spread before a stiff northeast trade wind and our little brig made the passage in something less than four days. The climate on the coast of Honduras in the month of December is extremely variable, a good deal of rain may be expected, but one gets rid of the excessive heat of the summer and fruit seems to be just as plentiful.

The omnipresent cocoanut tree is on all the keys, in fact you sight the cocoanut trees long before you see the low sand islands which are not very high above the ocean's crest, and the cocoanut milk, of which we drank a great deal, was a luxury we enjoyed about as much as any fruit we got. All along the coast, sometimes extending off shore many miles, are these little sand islands, or keys, as they are called. While one meets with some good islands in the vicinity of Balize, such as Bonnaca, Roatan, and Utilla. These islands have extensive forests of hard wood. They are, however, infested with poisonous fleas and insects, consequently all the inhabitants flee to the wind-swept keys for comfort. They procure their water from the main islands.

We made the coast in the vicinity of Ambergris Key, some forty miles north of Balize, and about three miles off shore and found them to be a very dangerous piece of coast, but Captain Soper seemed to like to sail around such places. We ran into the thickest of a group of shoals called the Triangles, and they might have been the Symplegades and we the crew of the Argo, so far as danger was concerned for Captain Soper acted on this occasion as though he was a descendant of old Jason, and was acting under the protection of the Gods of the Fabulous age.

It was a beautiful morning and the water and land looked inviting, and so we steered in among the reefs, keeping a good lookout at masthead and a leadsman in the main chains throwing the load. "A quarter less three" came from the man with the lead; but as we only draw nine feet of water aft and about eight feet forward, sixteen feet seemed pretty deep to our old Jason, the next two or three casts of the lead gave from sixteen to fourteen feet, and then the leadsman called out, "by the mark two." Captain Soper was on the brigs bow looking for a sand spot in which to drop anchor, when the writer remembers feeling a grating sensation and next thing we knew, we were aground on a coral reef.

We let go our anchor and dropped our sails and commenced to investigate our position, and found that for the present we were not in any great danger, for the coral was soft and there was at least two or three feet of fan-coral lying between us and the hard rocks, and so long as it did not slow and there was no sea running we had nothing to fear. We immediately got out a kedge anchor and lowered one of the boats and took the kedge out into deep water and dropped it, we then hove up our working anchor and taking the kedge-anchor line to the winch, we soon pulled the brig off into deep water.

After this we dropped anchor again and looked around to see what damage had been done. We found we were unharmed save from the loss of some of the copper on our keel. Then we commenced to fish and succeeded in catching quite a lots of fine fish, a rod codfish and two largo baracotas. The bottom was very pretty on this reef. It was covered with different colored sponges and long fans of coral. Little fish were darting around and hiding in the recesses of this charming marine forest. The water was so clear that the bottom could be plainly seen in ten fathoms.

Later in the day the wind began to rise and so we hove up our anchor and made sail and stood off shore for the night. Captain Soper seemed well satisfied with his adventure and but for the fact that we carried a ragged keel for the balance of the voyage, we were none the worse for having pounded a bed in the coral on the coast of Honduras.

Chapter Eight

Blackfish and More Fruit

In my last chapter we had just escaped "through the skin of our teeth" as the second mate remarked, leaving our bones on the dangerous Triangle Reef. After leaving that ground, we worked off to the eastward and for a week cruised very industriously in search of whales, but outside of killing half a dozen blackfish we got nothing. Our blackfish, however, furnished us with a little excitement.

One of them, a big fellow over fifteen feet long and weighing over three tons, came very near breaching into the first mates boat. Had he done so it would have been goodbye mate and crew to say nothing of spoiling the voyage. We were afterwards told of a case where three men were killed from a blackfish swamping a boat and we have no doubt the story was true.. We got quite a mass of oil from the blackfish, four or five barrels, I think. We used it for our lamps, the oil making a good light.

Sunday coming on we worked south towards Balize and again made the land, this time off the Northern Two Keys. They are low coral keys their shores fringed with cocoanut trees and tropical verdure. Some low huts were in sight, so we lowered a boat and went on shore. We found some white people. They claimed they were of English descent, but the relationship must have been remote several generations, anyway. These people were very glad to see us and invited us into their houses. But the inside looked less inviting if possible than the outside as we did not stop long. They treated us to cocoanut milk and a kind of liquor made from pineapples, which was white, very sweet and pleasant, and was capable of furnishing a first-class jag in a short time. The people wished for some reading matter, any old papers or magazine. We promised them some and two of the men followed us on board the brig and we gave them a big supply of what we had. They seemed delighted to get them.

We left Northern Two Keys and ran off shore for the night and next morning we ran down the coast further towards Mauger's Key. A heavy squall coming on, however, we were obliged to run off shore and shorten sail. It blowed very hard for twenty minutes or more and the rain left everything dripping. The land came out of the squall a sparkling mass of green, and the sky seemed a brighter blue than before.

About noon Goliah sent one of the crew aft with the statement that he Goliah, had got a needle in his thigh. He created an excitement. But all search for the needle proved hopeless. The second mate said, "You couldn't expect to find a needle in his nasty black hide," but Goliah insisted that he could feel the needle. However, he was around in a day or two and there was no bad effect apparent.

About noon, the weather improving we tacked inshore again and soon saw lighthouse reef, with a small lighthouse and near it the bleaching ribs of a vessel of some kind that had been wrecked there in years gone by. It is a curious fact that many times you will see wrecks in the West Indies right under the very tower of a lighthouse. Captain Soper explained that a number of heavily insured vessels were lost on those keys, as they afforded a safe landing place for a stranded crew, and that most of these wrecks were old vessels that had outlived their usefulness.

The next day being Christmas we had an extra good dinner on board the brig. The meal consisted of canned roast beef and a big dried apple duff with plenty of sugar sauce. We noticed the mates coming from dinner wiping their mouths with the back of their hands. It looked suspicious for a temperance vessel, and later in the day the old third mate said they had "spliced the main brace." Christmas was a blue day for most of us. The second mate was a downright ugly, but Captain Soper and the others were as usual kind to all.

I find the following in my journal of December 26th, 1873. "The weather was rough all last night and about eleven o'clock we parted the chain attached to the port foretopsail sheet, and one spell it looked as if we would split the sail before we could get it clewed up. But the watch rushed to the clewlines and buntlines and soon had the slatting chunk of canvass controlled. The Captain came on deck during the excitement and ordered the sail furled, which was done, and the balance of the night we lay under a foretopmast staysail and the main trysail."

In the morning we made sail and run in towards lighthouse reef, and soon after taking on board a pilot, who agreed to anchor us in good safe place for a big piece of pork, we dropped our mud hook on the southeast side of the island in three fathoms of water. After furling our sails and paying off the pilot with a couple dollars worth of salt pork, Captain Soper ordered the waist boat lowered and went ashore to see the key and the lighthouse.

Eight. Blackfish and More Fruit

The key was a little coral island with an area of about ten acres, and bounded on the North by an ugly reef on which lay the bones of the vessel mentioned above. The key was infested with an army of rats, they having swam ashore from the wreck and they seemed to be at home there. There must have been quite a fight between the inhabitants of the key and the rats. The residents were the lighthouse keeper and his family and two or three families of fishermen who were fishing for the Balize market.

The lighthouse, an English structure, bearing the trademark of a Birmingham firm, stood on the extreme northern end of the island.

The light was made with cocoanut oil. The keeper showed us a large vat of the oil. All we remember about the oil was that it was cream colored and stunk enough to warn any mariner off shore if he ever got a whiff of it. The glass in the tower, a thick one, was cracked in several places. We asked the keeper what cracked it. He told us that during hard storms sea birds would fly for the light and striking it with great force not only killed themselves but broke the glass.

A big bird weighing ten pounds or more flying at a speed of fifty or sixty miles an hour would strike with great force. The keeper told us that previous to the installation of his light wrecks had been frequent, the route to Balize passing the island, and he hinted that some of the families on the island had reaped a rich harvest every time a vessel came ashore as a ship never got off. His salary of four pounds a month would not have appealed to us very much, in fact we would rather go whaling. Looking over the island we found leaning against one of the houses a figure-head of some wrecked ship. It had been carved to represent some beautiful lady, but some artist on the island had given her cheeks a coat of red mineral paint and her beauty was a thing of the past.

After we went aboard the Captain let the third mate and a lot of the other boys go ashore. They tarried quite late and when they came aboard the ship they were very noisy and the old third mate was pretty well soused. They had been drinking something stronger than cocoanut milk, that was plain. But they had no money! We found out next day that they had traded their shirts for a bottle of Holland gin. "Where there's a will there's a way."

[Editor's Note: On Christmas the log noted that the crew went ashore and found "Cocoanuts in abundance and got a boatload and gathered some shells and came on board so ends another Christmas."]

Prior to Christmas the crew repaired the water closet or toilet and "saw queer looking fish in a white water."

Chapter Nine

Cruising in the Tracks of Columbus

The entrance to Balize is guarded by a large number of little keys which extend well out to sea. So numerous are they that the early settlers have been obliged to have recourse to the points of the compass to find names for all of them. Such names as Half Moon Key, Triangles, Alacroes, Grover, Hat, Ambergris and the euphonious Hog are here, and then there is Northeast, Southeast, Southwest, etc.

We are in historical waters. The grand old captain from Genoa, Columbus, must have sailed over these same waters in which we are now cruising as early as the year 1502. Who knows but what he visited some of these little keys himself. He may have slaked his thirst with cocoanut milk as we are doing, for cocoanut trees no doubt were there as now. What must his thoughts have been while cruising here. In those days they knew nothing of the new continent.

It had just been discovered and the advance of exploration was being worked out with the upmost caution and the greatest of difficulty. History tells us that it took Columbus from October 12th, 1492, when he first made Wallings Island, until August 1st, 1498, nearly six years, before he set foot on the South American soil, somewhere in the Gulf of Paria, near the mouth of the Orinoca River.

The cupidity of Columbus and his crew held them enthralled almost among these islands and those that lay to the east of them. In the winter of 1494 Columbus sent to Spain four of his ships loaded with the fruits of his exploration. Gold, precious stones, fruits and five-hundred Indian captives to be sold as slaves. The compassionate Queen Isabella ordered the slaves returned to their native lands, which was done, but their lives on their return were rendered burdensome by their Spanish taskmasters.

On his fourth and last voyage, in August 1502, Columbus discovered

Nine. Cruising in the Tracks of Columbus

and landed upon Cape Honduras. He was then in search of the strait which he supposed existed to the southwest of Cuba. The extraordinary lapse of time which existed between his discovery of San Salvadore and the contiguous coast of America, will always be a wonder to historical students. It was not until 1508, sixteen years after Columbus's great find, that Sebastain de Ocampo discovered that Cuba was an island by sailing around it. Columbus seems to have died in May, 1506, ignorant of that fact.

I have often wondered if the visionary Ponce de Leon was with Columbus on the latter's last voyage, when he discovered Honduras. If so the contrast which he met with in 1521 when he landed for the last time on the coast of Florida, must have been great indeed. Here in Honduras was all peace and quiet. Perpetual Summer and a peaceful race to greet the strangers. In Forida he was met by savages and had a constant battle with them and finally received the wound that killed him. Ponce de Leon had at last found the "spring of life" for which he had so long sought.

This is a land to dream in. Nature has done so much for its people. Left alone to your own thoughts you cannot help but dream if you have the love of nature in your soul. Like the State of Washington, this is an "Evergreen country." Constant verdure is here in and out of season. Even the coral reefs into which we ruthlessly drop our rusty old anchor are green on their tops with a rank sea moss or vegetation. The waters which impinge the coast are of a transparent green, and so clear are they that the bottom can be readily seen in calm weather where the water is ten fathoms deep.

Myriads of fish dart in and out among the many colored sponges and coral ferns that cover the sea floor, and great conch shells lay half hidden in beds of branching coral. Growing almost down to the waters edge are the cocoanut trees, a little further back among the houses of the natives, are the banana and plantain trees and a short distance further back are the pineapple, yam and sweet potato. The coffee tree is also here, also the cacao, better known as the chocolate.

This is the tramp's paradise, but there are no tramps. Nature encourages indolence here but the tramp is missing. It requires so little labor to gain a living that one wonders that people here have any ambition at all, and they don't manifest much. I once heard a West Indian minister advance the argument that the Garden of Eden was located in the tropics, and that Adam and Eve were driven north of the 35th parallel of latitude for their sins. A visit to Honduras will convince one that they can live here on as little and with as little effort as any place in the known world.

Chapter Ten

Cruising Among the Islands

December 29th, 1873.
Came in rainy and disagreeable. A strong northeast gale is blowing. We waited until near nine o'clock for it to clear and it having moderated some we hove up our anchor and started. We ran along to the windward of Grover's reef, a dangerous coral reef some fifteen miles long. The sea was curling and dashing over its top; presenting a long line of surf as far as the eye could see. We could not help thinking how easy it would be for our vessel to "miss stays" and land broadside in this reef. With such a breeze as was blowing this morning the salvage would have been small and the insurance large and the crew, oh, the Devil take them, what right had they in such a place.

This reef was said to be noted for turtle and towards noon, the wind having moderated sufficient, Mr. Hussey lowered his boat and pulled in for the reef and hunted around in the rain for over two hours. They finally came on board drenched with the rain and disgusted. They got no turtles and did not see one. We ran the length of the key and then hauling more to the wind headed over towards Southeast Key where we dropped anchor early in the afternoon. Our anchorage was three quarters of a mile from the beach on the west side of the island, in four fathoms of water. There is a good anchorage here for a vessel of any size. The shores of this key are fringed with cocoanut trees and there are a few people living here. The fruit boats from New Orleans come here every month for fruit and cocoanuts and the people do quite a business with the "Yankees" as they call us.

The next morning it was rough again. Along about nine o'clock we saw coming through the rain and squalls a little boat. It was a pilot from the Middle key. He had pulled three miles or more through the storm and rough water and was drenched. I don't think I have ever saw as mad a man in many a day as he was when he found out that we were not going to Balize,

Ten. Cruising Among the Islands

and so had no use for him. He gave a noisy exhibition of West Indian profanity. The crew laughed at him, and that made him madder, and he refused to come on board, although Captain Soper was very polite and invited him to do so. He pulled off in disgust, and we saw him battling with the waves until he was out of sight and nearly over to his key. This was the first time and the only time that a West Indian pilot refused to "come on board and be warmed up." The old grouch missed getting a good big piece of salt pork.

Chapter Eleven

Southwest Key and One Comfortable Home

December 31st arrived with a heavy rain, but towards noon it abated and during the balance of the day the weather was all that could be wished. In the afternoon the Captain and myself went ashore on Southwest key. There were quite a number of families living on this key and we spent a pleasant afternoon with one of the most prosperous of them. He was of English descent, but he and his father before him had lived here or in Balize. A good many years ago England sent many of her convicts to her West Indian possessions, and their grand-children can now be found here. Of course this is not talked about and they are not to be blamed for it. This family I have in mind lived in a house whose walls were of ironwood, but it had a thatched roof, and its owner claimed it was a tight roof, too.

The floor was of hardwood, too, and was scrubbed scrupulously clean. It was a single story house all one room divided by low walls that did not reach to the ceiling. 'The beds were surrounded with mosquito netting, indispensable here I should think. They had pictures on the walls and a Singer sewing machine. All their cooking was in a cook house some fifty feet from the house.

They used coal oil lamps and had many American conveniences. They had quite a nice library and there was a little whatnot made of spools, such as we were making in the States at that time, and on this Whatnot was some of the most beautiful shells and coral I ever saw. At the doorway lay a sorry looking dog, languidly fighting the fleas that were worrying him. We were treated to fruit and cocoanut milk and on leaving the house on our walk down to our boat surprised a young woman in a pair of pants climbing down out of a tall cocoanut tree. It was the first time I had ever seen anyone gathering cocoanuts, so I was not to blame if I did stare at that girl. We were told that it was a common practice for the girls to gather cocoanuts whenever they wanted them, and it is quite a trick too, for some of the trees are forty feet to their feathery tops where the nuts grow.

Chapter Twelve

New Year's, 1874

New Year's dawned with a strong wind from the North, but by ten o'clock it had moderated and we took our anchor, made sail and stood across the bay. We saw no whales or signs of them, so we came back again in the afternoon and anchored in our old mooring place. We mended the jib and rove off new foretopsail clewlines. After supper the Port watch went on shore and spent the evening.

January 2d. It was rainy and pleasant by turns today. Early in the morning we hove anchor and with the wind for the northeast run off to leaward of Southwest key and then hauled to the wind and stood along the south side of Five keys. Later in the day, the wind hauling more to the westward, we stood up along Grover's key and before sundown sighted Long key, bearing north-north-east about ten miles distant. This is the first night for a week that we have not anchored.

January 3d. The weather and wind very pleasant. Cruised in sight off Northeast key. One sail in sight, a schooner bound along into Balize.

January 4th and 5th were beautiful days and we cruised along the South side of Hat key, and anchored at night south of that key. On the morning of the 5th when we went on deck, we found we were laying in water as smooth as a millpond, not a ripple and clear as glass. Nine fathoms of' water and our anchor plain in sight in its coral bed. The bottom was a kaleidoscopic mass of sponges and sea verdure. The long cable leading from the brig to the anchor looked no larger than a watch chain and the anchor resembled a charm. Hosts of fish of different varieties darted around under the brig, sometimes turning on their sides to look at us. We watched them for a long time and endeavored to catch some but they would not be caught; no matter how adroitly we baited the hooks and how gently we lowered them into the deep they would not be deceived and at last exasperated with our luck on board, we lowered a boat and sent her in by the reefs and after a good deal of labor succeeded in spearing some of the bigger ones who came near enough.

Part I—"The Voyage of the *F.H. Moore*"

About ten o'clock a breeze springing up we hove up our anchor and started over toward Key Bokell, which we reached in the evening. It has two lights, one a white the other red, hoisted one above the other on a staff.

The lights, one can be seen about twelve miles. About twelve o'clock tonight there was a school of blackfish playing around the brig.

On the morning of the 6th we cruised off Key Bokell and in the evening dropped anchor off the southwest side of that key in: five fathoms of water near the light poll. The 7th was very windy, so we lay at anchor all day. Last night it came on to blow very hard and about midnight we called all hands and dropped a second anchor and made all ready for a gale as the barometer was falling. But it did not come and all our preparation was for naught. On the 8th we again took our anchor and making sail worked to the eastward past Long Key and that evening anchored off Half Moon Key. This afternoon a pilot came on board and reported having seen whales two weeks before off shore some thirty miles to the eastward in the vicinity of Bonnaca. He got a piece of pork for his trouble in coming off to tell us. We cruised around Half Moon Key until the 17th and not seeing any whales we started for Bonnaca, and Roatan. We sighted both of those islands late in the afternoon of' the 17th. Sunday the 18th was a fine day with light northeast trade winds which were cool and refreshing. The island of Roatan is in sight. There is a strong southwesterly current there, about two knots an hour. We have had a visitor for the past few days, a little land bird probably blown off from the island and tired out, he seems much at home with us. Blackfish in sight this evening.

January 19th and 20th were fine days and we cruised to the westward and on the latter day were again in the vicinity of Northern Two Keys We steered to the westward of them and anchored off Long Key. In the afternoon we sent a boat. to the Lighthouse Reef to see if we could get some copper sheating for the brigs bottom, which is getting ragged. We found we could get none, but another boat which went ashore later for cocoanuts returned loaded. Tomorrow if the wind is favorable we shall return to Roatan.

January 21st was rather a bad day for cruising so we remained at anchor all day. In the morning we sent down the foresail for repairs and worked on it off and on all day, finishing it late in the afternoon. About nine o'clock a little schooner named the Cereta, of Balize, came to anchor near us. Captain Soper sent a boat aboard of her and invited her Captain to dine with him. His reason for doing so was that he not only wanted company, but he wanted to find out from the old West Indian if he had seen any whales lately. The Cereta had left Balize the evening before and had on board some

Twelve. New Year's, 1874

seventy men, who with their families were bound for the island of Romain, a port down the coast, to cut mahogany timber.

It had been raining most of the night and as the Cereta was a small one, all those who were obliged to stay on deck were drenched and in a demoralized condition. Some of them were pretty well filled with liquor, too. The cabin of the Cereta was a little hole 12 × 14 feet square. This little dark hold had been appropriate[d] for the women and children, while the Cereta's hold was used for the men of the party. Conditions below were shocking and we wondered if the hold of a slaver could have been worse.

It was a Bedlam of filth and noise, stench and misery. The men were "lumber jacks" in every sense of the word. On the deck pure air rendered things much better, but all of the goods and chattels of these poor people were soaked by the rain and were a hard looking lot of junk. The families of some of our Pacific coast lumbermen have some pretty tough outfits to look at, but these Honduras lumbermen were the limit. These people were bound to a fever infested coast and the remuneration they were to receive for their work was $10.00 a month and their board.

Verily, whaling is a Paradise in comparison. After dinner the skipper went on board his vessel and was watched him heave his anchor, set sail and depart with his seasick cargo. We all gathered at the rail and gave them three hearty Yankee cheers as they sailed away. I have often wondered how they fared when they got to their destination. The next day we bent our foresail and hoisting out anchor started for Roatan, which we reached on the 24th. We took a route further south than on the previous passage and sighted the big island of Utilla. This is quite a high island, a dozen miles to the Southwest of Roatan.

The morning of the 28th dawned squally and disagreeable. The Northeast trade wind was very much like a gale and we found ourselves to leaward of Roatan and Utilla, and as we wished to work windward of a group of islands called the Hog Islands, we set a reefed mainsail and drove the little brig through the water like a mad vessel. The water boiled and seethed under our stern and the little pilot fish which were our companions and had been for several weeks, swam close to the rudder and swam with desperation to keep up with us. The bow of the brig was drenched with spray and all hands were drenched also. Late in the afternoon we were amongst the keys again and were grouping our way to find an anchorage, where we could lay until the weather moderated. The channel was narrow and the water shallow and the man with the leadline in the main chains was kept busy.

Part I—"The Voyage of the *F.H. Moore*"

After two hours of worrysome work we dropped anchor and furled our sails. It was one of the hardest days work of the voyage, but we made it all right. No sooner was our anchor on bottom than we were surrounded by a fleet of canoes. They were loaded with fruit of all kinds, some delicious pineapples and great bunches of bananas and soon a lively traffic was going on and the trading was kept up long after the sun set. This is one of the prettiest spots in which we have anchored.

To windward as far as the eye can see are little keys covered with cocoanut and trees of hardwood varieties. We catch a fragrant breath of the sandalwood and there is the sound of running water from a little stream which is running down from the hills into the ocean. It is a fine place to fill our casks with good drinking water. The tradewind is rustling the feathery tops of the cocoanut trees, and we can hear the dull boom of the surf on the windward side of the island. We sat on deck long after everything was quiet and the trading was over and all the lights on shore had been doused and thought of New England and compared its rocky coast and this tropical paradise. The watch on deck got to telling stories and Sam, the boatsteerer, told the following one.

Editor's Note

As the *F.H. Moore* continued to hunt whales with little success the log entries became shorter and more routine. A sense of boredom crept in as on February 20th, the entry read "saw nothing as usual." The tension built up among the crew.

On March 2 the log noted "a pitch battle between two Niggers," following the habit that black members of the crew did not even merit the mention of their names in the log book.

On Saturday, March 8th the log noted that "George Watkins and Elie Flemings had a dispute the latter drew a knife but was arrested before any damage."

By Monday Captain Soper apparently had had enough. "This was a day of trials among the crew," the log noted. The assembled crew were "read Sections of the Articles also some of the customs of the Ship." The two combatants were let out of irons, but another crew member was called before the masthead and punished for "skylarking on the Sabbath" and another was punished for drinking.

Chapter Thirteen

Sam the Boatsteerer's Story

Story telling on board ship passes many a pleasant hour. Of course there are all kinds of stories told. They generally recall some adventure of the narrator or his friends. The following is a true story. It is not a very exciting one, tis true, but it illustrates how close a call some crews have and still follow the sea.

"It's always darkest just before dawn' is an old, old saying, and I have often thought it true, especially after a long and tedious night of watching on ship board" said Sam the boatsteerer.

"When just before day the very stars seem to put out their lights, the moon sinks below the horizon and everything on board the ship and at sea is shrouded in blackness. But daylight begins to break and the impatient watchers rub their sleepy eyes and look about the [word omitted in text] and is it any wonder if they naturally scan the horizon for some passing sail? Often have I known two weeks to pass without seeing a sail, and how lonesome we were. There is no solitude so pronounced as that of the ocean, unless it is that of the coast of Africa about 200 miles South of the Cape Verde Islands.

"The wind had left us during the night and when morning broke the sea looked like a huge sheet of blue glass. There was the long, continuous ocean swell, and our vessel's head rose and fell on it, but not a ripple broke against her side. The sails hung almost motionless against the mast. The silence was profound. But all of a sudden the east commences to lighten, and day is breaking. But long before the sun rose like a ball of fire out of the ocean, all the watch on deck were on the alert to find out if there were any sails in sight.

"There was a sail in sight this morning and only a mile or so distant at that; it was a little sloop of about twenty tons; such a little craft to venture so far away from land. We noticed that she has no sail set, but as it is a "dead calm" we think nothing of that. Our heavy spyglass revealed her crew pumping and a stream of water running out of her scuppers. But that was nothing strange, for it was the custom on many vessels to pump ship and to scrub decks early in the morning. But soon our little neighbor hoists his flag. We watch it as it ascends, but suddenly it stops and hangs at half-mast. It was the red and yellow flag of a Spanish merchantman, and he was in distress.

"Our Captain was called and after watching her for some time through his glass he said 'Thompson, that little fellow is in trouble and needs some assistance, lower your boat and pull over to him and find out what ails him. If he has yellow fever on board, don't go near him, for I don't want this crew exposed and he will have to get along as best he can.'

"With these instructions we lowered the boat and pulled for the sloop. As we drew near we could see what the trouble was with her. She had sprung a leak and was slowly sinking in spite of the heroic efforts of her crew to keep her afloat. Pulling along side of her we find that the pumps are not working well. The leather is worn off the pump boxes and they have no more on board of the vessel with which to repair them. The water is gaining on them slowly and the little sloop is settling slowly but surely. It is a question of an hour before she will be on her dive for Davy Jones locker. We must work quick if we would save the little craft, and we do.

"Telling the old Spaniard that we will go on board and get some leather and come back and help him we jump in our boat and pull back for our ship. We pulled the mile between the two vessels in quick time I tell you, and were soon pulling back to the sinking sloop, this time with two more boats coming to help pump her out.

"Eighteen fresh, strong young fellows ready for an adventure. When we reached the sloop we found one of the crew had dropped to the deck from exhaustion and the balance of her crew were clamoring to the old Spaniard to desert the sloop and seek refuge on board our ship. But we soon had his pumps repaired, our fresh crew of men pumped like good fellows and some went into the hold and bailed with buckets and before noon we had the leak safely above water and our carpenter went over the side and stopped it.

"The little sloop had been trading on the coast of Africa and her cargo consisted principally of cuttlefish bone. She was on her way to a port in the Mediterranean. You never saw a happier lot of men, especially the old skipper. He hugged us all. He didn't have much on his little craft, but was wiling give us all he had. We finally were obliged to accept a lot of fowles.

Thirteen. Sam the Boatsteerer's Story

He would have killed the 'fatted calf' for us. Of course our captain would accept no pay, but he did accept a bottle of wine which the Spaniard thrust into his hands as he climbed down into his boat.

"Later in the afternoon when the wind sprang up we saw him hoist his sails and steer to the north, all time raising and dipping his flag. We did the same and that was the way we parted, never to meet again.

"He had a close call. But life on the ocean is made up of many such chances. And brave men take them and forget all about them."

Chapter Fourteen

[Untitled]

To leaward of our anchorage were low coral reefs, over which the sea was constantly seething and swashing. In places one could see open spots were there was blue water, denoting greater depth of water and a passage through the reef. The next day Captain Soper and myself accompanied by a boatscrew of men, sounded these openings, but in every case found the current setting so strong through them that it would have been hazardous to have attempted to make a passage through them with the brig. So Captain Soper was at last forced to the conclusion that he must depart the same way he entered the harbor. We made some repairs to the brig's bottom and purchased two chords of ironwood for our cook stove, took off twenty-five barrels of fresh water and mended our sails before sailing.

While in the boat Captain Soper visited several of the little keys and hunting amongst the coral reefs I found some very pretty shells. The next day we took our departure from Hog Islands and having worked through the dangerous passage once more, stood over in the direction of Bonnaca. We noticed a vessel lying at anchor there and we intended to have sent a boat on board of her, but it came off squally and we stood off shore for the night. The next day, the 28th we spoke [to] a pilotboat from Bonnaca and her skipper reported having seen whales a week previous. He also stated they were big whales and we come to the conclusion they might. be sperm whales.

January 29th. A fine day and we spent the time cruising around to the north-east of Bonacca, and towards evening we ran in towards the settlement and sent a boat on board of a schooner lying at anchor and taking on fruit. She was bound for New Orleans with bananas and cocoanuts and we availed ourselves of the opportunity to send letters home by her. We then returned to the brig with our boat and stood off shore for the night.

January 30th. This has been an exciting day. This morning we sent down the middle-staysail for repairs and this afternoon at 3:30 o'clock, Mr.

Fourteen. [Untitled]

Huzzey, the 1st mate, raised a school of sperm whales. We lowered all three boats and after a short time the starboard boat struck and then the port boat and we got two small whales. The whales did not seem to be going very fast and are probably at home. We got the whales alongside before dark and made fast for the night will cut them in tomorrow.

January 31st. Was a fine day. This morning we started cutting early and by 10 o'clock had both whales on deck and all hands cutting the blubber into horsepieces for the mincing machine. We had the heads cleared off by four and then started the fires under the pots and commenced to boil out. The island of Bonnaco is in sight tonight bearing southwest.

Sunday and Monday February 1st and 2d, 1874, were nice days and we finished boiling out the oil and stowed it down. There was 28 barrels. We are cruising where we saw our whales last Friday.

Tuesday, Feb. 3rd. Fine weather today. This morning we cleaned the oil from the decks and rails and about five o'clock Mr. Streeter raised whales again. We lowered and Mr. Huzzey got fast to a little whale and the rest disappeared as if by magic. We did not see them again. They must have gone a long ways underwater. We got the little fellow alongside by dark, put up the cutting gear and will cut him in tomorrow morning. The whales were working over towards Roatan Island.

Wednesday, Feb. 4th. Commenced to cut our whale at 5:30 and had him all aboard before breakfast. Cleared the deck and made sail and started after the whales, that is we started in the direction we thought they would go. Sent down the cutting gear and started to boil tonight. This evening out [our?] cutting stage got loose, and fell into the water and as the brig was running about five miles an hour at that time the stage was wrecked before we could stop and pick it up. Bonacca and Roaton Islands both in sight.

Sunday, Feb. 8th. It blowed hard all day from the northwest and we made for Bonnaca and came to anchor. There were several schooners, fruiters, at anchor and we took the opportunity to send letters home. Monday the weather continuing boisterous, we laid at anchor and traded flour, beef and pork with the natives for fruit and cocoanuts. Late in the afternoon the wind abating some we took our anchor and got under way. [We] Are going back to the place where we got our whales in hopes of getting some more.

Sunday, March 8th. For the past month we have been cruising around Bonnaca and as far east as Swan Islands. We did not land at the latter, however. The weather has been fine most of the time. We saw no whales but saw blackfish and caught a couple of porpoises. A porpoise on board of a whaler is much preferred to a hog. The meat is quite dark and when chopped fine and mixed with fat pork with a dash of sage or summer savory and fried makes an excellent sausage meat, or you might call it hamburger. The brains of a porpoise are fine eating. There are about five or six pounds of them and when mixed with cracker crumbs and fried crisp, they are excellent eating.

This afternoon one of the crew and the little black cook got into a fight over some trivial matter. The first thing we knew the big black fellow named Eli had his knife out and was going to carve the little fellow who armed with a belaying pin was as full of fight as a banty rooster. Tucker and I. separated them and the captain coming on deck, ordered both to be handcuffed and put down in the hold. The next day they were both very penitent and after giving them a good lecture Captain Soper ordered the hand-cuffs off and the pugilists were ready to be good again.

Thursday, March 12th. I find this entry in my journal. A hot day. Quite a number of porpoises in sight. About eleven o'clock this evening as the heat was something tropical below in my room I thought I would try and sleep on deck in my boat. I took a pillow and blanket and in getting into the boat both got away from me and fell overboard and as we were going ahead quite fast I lost both. "Bad luck to Biddie McGee."

Tuesday, March 17th. Came with a strong breeze and continued squally. Early this morning we raise sperm whales but they were going to windward fast. We lowered the boats, however, and after a hard pull for an hour gave up the chase and came on board. The whales were going fast to windward, probably something had frightened them.

Thursday, March 19. We saw the whales again today. Early this morning between 8 and 9, Horace Pease, Mr. Huzzey's boatsteerer, raised a school of cows and calfs. The weather was fine and we had to work very carefully not to frighten them and it was not until one o'clock before the mate go fast, then the other two boats fastened and before night we had four dead whales alongside. We got two in our boat. Mr. Huzzey killed two also but lost one. The whales laid around like logs.

There must have been two dozens of them. I counted six or seven little calves. Mr. Huzzey's whale is the largest, a 40-barrel bull, I think.

Friday March 20th. Whales in sight all day, but we did not lower the boats for as we had our hands all full saving the four caught yesterday. It

Fourteen. [Untitled]

took us all day to cut them in and as it was we had to leave the head of the biggest one overboard until tomorrow. The weather is fine. Hope the whales will stick around until Sunday. We will then be in shape to lower for them again.

Saturday March 21st. Came in with a strong breeze and quite a heavy squall. Early this morning we finished taking in the whales head and commenced to boil the body. About noon we raised whales again. They seem at home and were going to leaward slowly. We have our hands full with the blubber on hand as the weather was squally we did not lower the boats. Quite a heavy sea running.

Wednesday, March 25th. Have seen no more of the whales. We have been busy boiling the blubber and stowing down the oil in the hold. We got 65 barrels of oil from our catch of last Thursday. We have the decks clean again, our boats and lines all ready and would like to see a school coming alone again. This catch makes 110 barrels so far this season.

We cruised about Roatan and Bonacca until April and then worked our way north to the Caiman Islands and dropped anchor in Georgetown, Great Cayman. We expected to get letters here from home but were disappointed so after buying a supply of yams and sweet potatoes we hove up our anchor, made sail and sailed back to Bonacca. We cruised here until April 30th, when we again saw whales and were lucky enough to catch five of them in one day. They were small whales however, and only made about fifty barrels of oil all told. We have now over 200 barrels taken since leaving home.

On Tuesday, May 19th, not finding any more whales, and the brigs bottom becoming very foul from cruising about in the warm, tropical waters, our captain decided to go to Key West, Florida, and discharge his oil and ship it home, dock the brig and repair her bottom, so two days later we bid good bye to Bonacca and steered north for the Gulf of Mexico.

On May 22d, we were off Cozumel Island on the coast of Yacatan, and saw a wreck bleaching on the white sand beach of the island. We sent a boat in to visit her and found a man in charge of her. She was a Spanish bark named the "Progresso" and had been wrecked about six months before our visit. The old fellow in charge of her sold us some rigging and a small block, in fact that was about all he had that we wanted.

Cozumel is a long low sand island lying on the southeastern edge of the Courtouch Bank. There are no inhabitants on the island, but there's a very good anchorage on the west side of the island in from 10 to 13 fathoms of water half a mile from the shore. Before you strike soundings on the edge of the bank, you will have to cross a strong tide rip, running north and south, which has a three-knot current and should not be attempted in light winds.

Leaving Cozumel we worked north and east towards Cape San Antonia, at the extreme western end of Cuba. The wind was strong from the eastward for several days and we had to beat up against it and owing to the bad condition of the brig's bottom made slow progress. But on May 27th, we passed San Antonia and along the northern coast of Cuba, which looked very green and pretty, and later in the day sighted Dry Tortugas, seventy miles west of Key West. Tortugas is nothing but a low sand key, and you see the great granite Fort Jefferson sticking out of the water long before you see the land.

The fort was built in the days before big rifled guns were invented and was considered of little good even at the time of the Civil war, but were used during that war as a government prison and quarantine station. Several of the prisoners connected with the assassination of President Lincoln were confined here.

Thursday May 28, 1874. Comes in with a fine breeze from the Westward and about seven o'clock we sighted Sand Key lighthouse and at 11:30 we came to an anchor in the harbor of Key West, having taken a pilot about an hour earlier. There is a large fleet of naval vessels here, but. there are few coasters. The big fleet of war ships has been here off and on since last fall, when it looked for a time as if the United States and Spain would clash over the shooting of Captain Joseph Frye and the crew of the filabuster Virginius at Santiago de Cuba.

That occurred on November 7, 1873, when Captain Frye and 36 of his crew were backed up against a stonewall and shot and the next day 12 more of the passengers of the ill fated steamer were also shot. The murder of the Americans precipitated the revolt against the Spanish in this country, and, presto, the whole American navy was dispatched to Cuba and Key West. After considerable diplomatic fireworks the Virginius and the balance of her crew were surrendered to this government. The Virginius was a filibuster, pure and simple, and had no right to fly the American flag. Her crew, passengers and cargo of ammunition was designed to help the Cuban revolutionists.

The fleet at anchor consisted of the big monitors Dictator, Canonicus. Ajack, and the wooden frigate Worcester, the latter the flagship of the fleet.

Fourteen. [Untitled]

Then there were several smaller craft. We shall break out oil tomorrow, cooper the casks and ship it by steamer to New York.

Friday, May 29th. We have been getting our oil out and ready to land today and have passed a busy day. Captain sent off a couple of barrels of Irish potatoes this morning, the first we have had since we ate up our supply on leaving Boston. They tasted good and ought to, for they cost $7.00 a barrel. Provisions or all kinds are very dear here, especially fresh meat and vegetables.

Saturday May 30th. This morning we hauled in alongside the dock and commenced to roll our oil. Two mail steamers came in today and I sent a letter home to mother. We have all the oil [on] shore this evening.

Sunday, May 31st. A lovely day although a little warm. We had a lot of visitors aboard the brig today. The whaler Moore is as much of a curiosity to people here as one of the warships and the crew has been busy all day explaining about our whaling gear. This evening it is very hot and the mosquitos are something terrible.

Monday, June 1st. Today we have been taking everything heavy out of the brig and landing it on the wharf. We took out all our stone ballast, quite a job, lowered all our boats and will haul down to the marine railway tomorrow. Very hot today and we have all suffered from the heat. Mr. Tucker the second mate left us today. He will return to his home in New Bedford.

Tuesday and Wednesday, June 2 and 3d. We put in the time getting ready to pull the brig out on the marine railway. It is a small affair made for the small fishing smacks and we had to lighten up the brig as much as possible, but tomorrow we expect to be able to haul her out. The weather is fearful hot and the mosquitos at night are troublesome. I have a big piece of mosquito netting hung in front of my bunk and nights I lay and swelter and listen to the little devils singing and fight what get in at me, but I manage to get some sleep. I shall be glad, however, when we get to sea again.

Thursday, June 4th. Yesterday Charlie Tucker, our second mate left us and took a steamer for New York and thence to New Bedford, his home. Tucker was good officer but not much of a whaleman and he and Mr. Huzzey have had several tiffs, and the little brig Moore wasn't big enough to hold the two of them. I hated to see Charley go, but he can do better somewhere else. This morning we commenced to put the brig on the ways and it took the greater part of the day to accomplish the task. After she was safely shored up we all took a good look at her bottom. It shows the effects of her numerous contacts with the coral bottom off the coast of Honduras. All the shoeing is gone and the copper also.

The copper is especially ragged where we have been holding our whales. I suppose it is caused by the whales ribbing along side in rough weather, at all events the copper is thinner there than on other portions of her bottom excepting her keel where we scrubbed the coral reefs. The ship carpenters have gone to work on the brig and tomorrow we expect to give the Port watch some money and a run ashore.

Friday, June 5th. The carpenters worked overtime yesterday and today with a result that they finished putting on a new shoe on the brig's keel and cleaned her bottom and patched the copper where it was broken. One gang of men went ashore today. The captain gave them $5.00 apiece to spend. Hope they don't get drunk and come on board and make a row, they generally do. This morning I went on board the Worchester on an errand for the captain. Of course I didn't know much about man-of-war etiquette, and my attempt to land with my whaleboat on the starboard side of the big warship met with a curt command from the Marine on guard at that gangway to 'go around to port.' This I did and found the Marine on guard there more pleasant. He took my letter, called another Marine who took the letter and disappeared down in the bowels of the big ship. I waited, wondering what would happen.

Presently [a] Sargeant of the Marines came up out of the hold and asked me to follow him. I did so and went down into the next deck, the gun deck it was called. There was a fearful array of murderous looking guns and all that went with them and sitting around about the guns was the crew. The deck was spotless white and everything shining.

Then the Sargeant told one of the sailors to call the master-at-arms and in due time that worthy appeared. Then I learned for the first time the nature of my errand on board the flagship. Captain Soper had asked the Commodore to let him have some leg irons and handcuffs. The master-at-arms disappeared for a few minutes and when he returned he was accompanied by two sailors loaded down with handcuffs and leg irons and I was invited to take what I wanted.

I didn't know how many Captain Soper wanted but it seemed to more than half a dozen pair ought to be sufficient for a peaceful ship like the Moore, so I selected them and thanked the Sargeant for them. He was as generous as an Indian agent dealing out provisions to a Sioux squaw, and I picked up my load of hardware and climbed on deck and over the side

Fourteen. [Untitled]

into the boat with it, and heard one of the sailors "There must be mutiny aboard the whaler." I felt like a warden about to hang someone and we pushed off from the Worchster and rowed back to the Moore and you bet I didn't show my hardware to anyone but Mr. Huzzey when I got on board.

We never had any use for them, but the fact that they were handy, may have had a good effect on any of the crew who were inclined to act up, is highly probable.

Saturday, June 6, 1874. Today the carpenters finished work on the brig's bottom and about the tides turning right she will be ready for launching. The Starboard watch is ashore today enjoying themselves as best they can in the hot sun. The thermometer registered 90 in the shade. Sunday, June 7, 1874 has been a beautiful day. I took care of the brig today while the Captain and Mr Huzzey went to church.

In the afternoon I went to Sunday school and in the evening to church. After Sunday school I took a stroll over to the soldiers quarters and came back to the brig for tea.

Monday, June 8, 1874. Today we launched the brig and took her alongside the wharf and took aboard all her gear and other stuff we took out to lighten her. Later in the day we moved her up to Tiff's wharf. The little brig looks fine with her new paint on.

Tuesday, June 9, 1874. Today we took all of our oil casks together with wood, water and other provisions and stowed them in the hold "hung up and bilge free." Had to fill our casks with rain water. No other to be had here. When the steamer City of Wilmington arrives from New York today she brought out new second mate, William Huzzey. He is a son of our first mate. Today ten of our crew deserted, which makes us short indeed.

This evening we pulled out into the bay. Don't know how soon we shall be able to get away from here as a men are scare [as the men are scared]. If we fail to get the men back we will have to ship a new crew. This has been a hard day's work.

Wednesday, June 10, 1874. Today we have been stowing the provisions in the hold and cleaning the deck. This afternoon I made a visit to the double turret monitors "Dictator" and "Canonicus." They are two of the finest warships in the American navy.

Thursday, June 10 [11?] 1874. This morning we cleaned the brig some more and then sent up the top gallant and royal yards and bent the sails. We are getting ready to leave. Went ashore this afternoon and visited one of the big cigar factories, where they make the famous Key West cigars.

Part I—"The Voyage of the *F.H. Moore*"

There was over 500 men and girls employed in this establishment and the tobacco comes from Cuba. Millions of the famous "Havana" cigars come from this place and the finest brands bring high prices.

This establishment is a bust place. There is a reader employed here during the working hours and silence is the rule of the big house.

Sunday, June 14, 1874. Here we are, still waiting for a crew. Men scarce and none care to ship on the Moore, who has been given bad name by the fellows who deserted from her. We hear that some of the deserters stowed on board the New York steamer which left here four days ago, at all events the police have not been able to locate any of the missing men. Possible they are not trying very hard to catch them as Captain Soper refused to offer a reward for them. It rained hard today, so I put in this day writing letters home. Got one each from mother and brother Frank.

Wednesday, June 17, 1874. Today we finished shipping a new crew and this afternoon we got under way, when we were going out of the harbor a boat from the flagship came alongside, and halted us. An officer come [came] on board with a search warrant and detained us thirty minutes hunting for deserters from the navy which were reported to be hidden on board the brig. They did not find any, but when we got to sea the next day two naval sailors came on deck, and we didn't go back with them. They made up our full compliment in the forecastle. We needed them. Singularly providential.

Editor's Note

On March 17th the crew spotted a pod of whales, lowered the boats and set out after them. The boats chased them all day until dark. The crew "Came on board after a hard day's work and nothing to buy the child's frock. The damdest whaling ever done and ever to be remembered."

By April 8th, the ship put in at Georgetown, Grand Cayman Island. The ship recruited some new crew members and had trouble with some others. One crew member was put in irons but later released after he claimed he could not remember what he did but expressed regret. The Captain, the log noted, "guess he thinks him a good servant but a bad master."

Three days later on Sunday, April 12th, the crew "passed the day in reading the scriptures and other words." A week later the log writer

Fourteen. [Untitled]

said he went ashore and "had a splendid time generally." While the ship was laid up for repairs, the log noted "had two niggers cut by playing with knifes one of them bad."

On April 30th, the *F.H. Moore* killed five whales. The crew paused to cut and boil the oil but had time on May 3rd, to go ashore a nearby island where they "gathered shells on a beach." The ship passed a schooner from New Orleans that gave them newspapers and passed on the latest sightings of whales, "finding a home for the Moore."

On Thursday May 28, 1874, the anonymous writer finished his entry "had a talk with the Old Man and consented to give me my discharge."

Chapter Fifteen

The Captain's Wife

Among the flotsam and jetsam taken on board the *Moore* at Key West was the Captain's wife. Mrs. Soper had written to her husband that she wanted to share his fortunes on the brig and he at last consented to let her have a trial of sea life. There were no accommodations on board the Moore for her and we have since thought it an error of good judgment on the part of Captain Soper to have yielded to her entreaties to be allowed to finish the voyage with him, but she had her way and from the time we left Key West up to the day we were towed into Galveston a dismantled wreck, Mrs. Soper was one of us and shared our fortunes without complaint—that I ever heard.

She was a little, grey-haired, slim, woman, past sixty years of age, who hung to the railing whenever she was on deck, as if fearful that the wind or some lurch of the vessel might send her over the side into Davy Jones' locker. As I said before, the accommodations in the cabin were poor enough for the captain and us officers, but for a lady, and an old one at that, they were unfitted. However, she proved a very good sailor. She got over the sea sickness quickly and came out of the ordeal smiling, something few women do, and she quickly got acquainted and had a pleasant word for all. In fine weather she spent a large share of her time on deck, usually under an improvised awning near the man at the wheel, and in spite of orders to the contrary, she carried on a conversation with every sailor who went to the wheel. She soon learned the family history of all the boys in the forecastle and they found her companionable, even if they did get the brig "off her course" frequently when she held her "conversations" as they called her interviews.

I don't think this talking with the man at the wheel worried anybody but the old mate, Mr. Huzzey, and he was too polite to intrude at any time. Possibly to have done so would have been made a scene and Huzzey was old enough to know that when a woman has made up her mind to find out the social standing of a neighbor, it is time wasted to interfere, and sometimes it is positively dangerous. Mrs. Soper never made any complaints

Fifteen. The Captain's Wife

about the victuals, and I could not see that we fared any better after her arrival, but she ate her "wack" as the sailors say, and it must have agreed with her for she looked better after she had been on board a couple of months. She put in her time sewing, reading and making fancy work, and in time seemed very much at home.

There was some profanity floating about the decks of the *Moore* at times, but if Mrs. Soper heard it she certainly made no complaints to anyone but the Captain, and he made no complaints to the rest of us. So we drifted along as if there was no lady on board. Sometimes we missed Mrs. Soper from the breakfast table, but she usually ate with us. Taking all things into consideration she proved a good sailor.

After leaving Key West we steered to the westward and reaching latitude 24.22 North, longitude 85.40 West from Greenwich, we found our whaling ground and on Saturday June 20th, three days out from Key West, saw sperm whales. Mr. Streeter raised them at about 10:30 o'clock. We lowered all three boats and chased all day, but the whales were wild and we did not get near enough to strike one. We cruised about and on the 23rd, found them again about 35 miles northeast of where we saw them on the 20th. We lowered and the Starboard boat got a small one which made about ten barrels of oil.

On July 4th I find the following entry in my journal. "Fourth of July" commences with half a gale and accompanied by a drenching rain. At 3 p.m., the fore-topmast-staysail sheet parted and before we could haul the sail down it was in ribbons. The wind played Old Nick for about three hours and then slacked up a little to get a better hold or more wind, when it blowed still harder for several hours. It has slacked up considerable now. Bent a new staysail to hold the brig steadier. Porpoises in sight. A nasty day for the Glorious Fourth.

July 7, 1874. Commences with light winds from the northward. Today we repaired the old staysail. Hands employed picking oakum. Later finished repairing the old staysail and rebent it in place of the new one. This evening we spoke the steamer "City of Austin" and compared chronometers. So ends. Porpoises in sight.

July 8, 1874. Commences with fine weather and was fine all day with light winds from the eastward. Course steered S. by W. all day. This afternoon Mr. Streeter cut out a new boatsail for my boat. The other boats are making gaff-topsails for their masts. Porpoises in sight.

July 9, 1874. Commences with wind from the southwest.

Mr. Huzzey is busy repairing the tryworks, which leak under the pots.

Part I—"The Voyage of the *F.H. Moore*"

I have been busy working most of the day on my boatsail. A great many porpoises in sight. The ground looks lively. We are now cruising where we got a big whale last year.

July 10, 1874. Light breeze from the south. Seeing rather smokey for some reason. Crew employed knotting rope yarns to make spunyarn. This afternoon we fleeted the gally from the port side of the foremast of the portside of the after hatch to make it more convenient. Worked on my boatsail some Porpoises in sight. About the 3d this month we began to see a beautiful comet in the southern sky. It has been visible every night since. We see it a little after sundown and about midnight it fades from sight. Do not know the name of the comet.

The current where we are now cruising is very strong and that fact is most noticeable after a day of calm weather, when we find ourselves thirty or forty miles off position. Captain Soper had a curiosity to test these currents in a different way so I built him a small boat ten feet long, a sort of skiff, and ever day when it. is calm we put the little boat overboard and throw the log from it. Our first operation is to anchor the skiff. This we do by dropping a ten-pound deep sea lead and paying out about 50 fathoms of line. The lead act as an anchor, the effect being to turn the head of the boat in the opposite direction to the surface current, if there is one. We then reel out the log and on some occasions have noted as much as a two knot current.

Another feature out of common on this Gulf of Mexico whaling ground is the amount of oil on the surface of the sea at times. Captain Soper says it is caused by the oil seeping from the wells at the bottom of the ocean. At times the sea for miles about is covered with a very thin slick of coal oil. I know it is coal oil, for I have tasted it. It is most noticeable after a heavy storm. It seems strange that whales would be found in such waters, as one would not think that the squid, the food of the sperm whale, could live where so much coal oil existed, but the whales are here in large numbers, and furthermore, this has been considered an excellent whaling ground for many year, and the whales we have seen thus far this season seem very much at home. It is one of the strange sights of the sea and reminds me of that verse in the Bible:

> "They that go down to the sea in ships,
> That do business in great waters,
> These see the works of the Lord,
> And His wonders in the deep"

Fifteen. The Captain's Wife

EDITOR'S NOTE

The comet spotted by the crew was Comet Coggia. Unlike better known comets like Haley's Comet, the comet, first spotted by and named for a French astronomer, is non-periodic and may not return for a considerable length of time if ever. Comet Coggia was noted throughout the world and aroused fears of impending disasters.

Chapter Sixteen

[Untitled]

July 15th we saw whales and got three, one to each boat. Our whales made us 52 barrels of oil. We cruised around on this ground until September 1st before we saw any more whales, when in the afternoon of that day I raised a school of sperm whales and we lowered and we got two. These were fated to be the last whales taken, as there came up one of those terrible hurricanes for which the Gulf of Mexico is noted and the brig was dismasted and the voyage terminated. The following entries from my journal relate the closing days of the voyage of the F.H. Moore.

Wednesday, September 2, 1874. This morning we commenced to cut in the whales. The wind from the Eastward has freshened a great deal, so we were obliged to run before it to lessen the risk of losing our mainmast while cutting in the whales. Later at sundown we finished cutting in the whales but it has been a hard days work, and while we have both of the whales heads on board and lashed to the starboard rail and the blubber in the hold, it is too rough to work at cutting and boiling. Put the brig under a close-reefed topsail for the night. So ends.

Thursday, September 3, Lat. 25.35, Long. 91.31 west. Commences very rough. About one o'clock a. m. we began to scud the brig before the wind, and two hours later, or about three o'clock, we shipped a big sea over the stern and lost the Starboard boat, the sea swept forward and filled the main deck and the little brig rolled and staggered under her heavy deckload. But the deck soon cleared. When the sea came aboard there was a terrible crash, my handsome boat on which I had spent so much work, and which to me was most dear, as I had spent many hours in her since leaving Boston, was filled full of water, and the great weight thrown on her was too much for the davits and crains to bear up under and they were torn from the side of the brig like so much paper. I saw her go, but was too busy hanging on for dear life and struggling in the water to pay much attention to her. There was a flash of lightning a few seconds after the crash and looking astern I

Sixteen. [Untitled]

saw the boat and wreckage on top of a big wave which seemed to be rushing for us, then it was blacker than ink, and that was the last I saw of the Starboard boat. The first mates boat on the Port quarter, was lashed to the rail and was saved, but before the storm ended she was a wreck, her planking having been torn from her stern and keel. At daylight the wind had increased to such a gale that we took in the reefed topsail and with the exception of a staysail to steady her we were under bare poles. Later in the afternoon we were obliged to throw the heads of the whales overboard.

Still later—At midnight (12) it is blowing a fearful gale and the brig is leaking, having manned the pumps every hour, but thus far have had no trouble keeping her free.

Friday, September 4th. What a terrible night it has been. This morning at 2:30 we were running before the winds under bare poles, the brig making terrible weather of it. The great seas followed her fast seeming determined to overwhelm her. At times we would be riding on top of a great wave that fairly threw her about like a cork in a millrace. Then she would sink down into the trough of the sea and the waves ahead and astern seemed like great green mountains of water and we held our breath and waited to be caught by one of them and capsized. We were in the same plight of Gonzalo, in the "Tempest" and could say with him: "Now would I give a thousand furlongs of sea for an acre of barren ground; long heath, brown furze, anything: The will above be done, but I would fain die a dry death." The brig is acting very badly. I have sometimes thought if it had been possible for us to have carried a reefed topsail we might have made better weather of it. As it was she would not mind her wheel butt veered from one side to the other, and about 2:30 she broached to on the Port tack and went over on her side, her starboard rail was even with the water and the main deck full of water. We knew the brig could never ride out the gale under such conditions, as she would soon fill with water, and Captain Soper very reluctantly gave the order to cut the weather rigging, to let the masts go over the side.

We cut the lanyards of the weather rigging and the shrouds swung over to leaward. But the stout spruce masts held and it seemed that we must capsize. The waves did not break over us for the oil in the hold was leaking and the ocean in our vicinity was slick with it. I climbed up on top of the main gaff and tried to cut away the mainmast. The wind was blowing so hard that it deflected the axe and I could not strike twice in the same place. I had made probably half a dozen ineffectual strokes with the axe, when suddenly the foremast broke off at its step in the keelson and with a crash went over the side to starboard. I dropped my axe and jumped to the deck and made my way to the Port rail. I could hardly done so when the mainmast broke off above our heads, and about fifteen feet from the deck and fell over to Starboard also. The wreckage was floating alongside to leaward and it drifted under the counter, and every now and then the brig would settle down on top of the wreck heavily and we saw we must get rid of it or we should bump our stern out on it. The main boom and gaff were thrown down on the Starboard pump and that very necessary and useful adjunct was wrecked. But the worse aspect was the foremast. It had gone down through the deck and rail but fortunately it stopped at the planksheer.

The mast was about fifty feet long and hung outboard [a] full thirty-five feet, and every time it dipped into the sea, and that was every time she rolled, it acted as a lever and buckled up the deck on the Port side. It was therefore necessary to chop it off at the planksheer, no easy job, as the mast was two feet in diameter. We were not woodsmen and the job of standing on the deck with the gale blowing and the brig rolling and cutting off a timber of that size was no snap, and it took quite a while, but finally we did chew it off and then the brig rolled all the harder. The jib-boom broke off at the collar of the bowsprit and the stays had to be cleared. It was hard, dangerous work, but we were nerved up to most anything and took some big chances, and finally after two hours hard work we cleared the wreckage away and had the satisfaction of seeing it drift astern. We then had time to look about and discovered that the brig was dangerously full of water.

We had but one pump that would work and were obliged to divide the crew into three gangs and while one gang pumped, the other two took buckets and bailed for dear life. At last we got the best of the water and the brig, with the help of the leaking oil, which kept the seas from breaking over us made very good weather, although the wind blew something awful all day, but towards evening we could see that the wind was abating, although still very rough.

With the decks cleared, we had a chance to look about us.

Then we discovered that the brig was full of water. I was given charge of half a dozen men and told to go forward and bail the water from the forward hold. Up to that time we had had little chance to count noses, and it was only then that we missed Stephen, a big black West India fellow, one of the crew. We of course thought he had been swept overboard during the night, but when I climbed down into the forecastle, I became suddenly aware that Stephen was very much alive, but badly frightened.

Sixteen. [Untitled]

He was down on his knees praying. "Oh Lord, come take Stephen, Stephen's ready any time." Stephen was certainly badly frightened. He was the worst frightened human being I ever saw. I grabbed him by the shoulder and pulled the trembling chunk of black blubber to his feet. Come now that won't do, get up and take this bucket and help bail. We'll get out of this scrape if we work. But Stephan was obdurate."Mr. Williams, let me alone, I want to make my peace with God. We's all going to the bottom of the big sea." There was no time to waste on him. So I pushed a bucket into his hands and threatening to bang him over the head with another, I set him to work. A little encouragement and the companionship of the other boys soon cheered him up and he worked as hard or harder than the rest, and as we began to gain on the water he took on courage and proved of first class assistance. He told me afterwards that he was down on his knees when the masts went over the side, and he no doubt thinks to this day that his praying saved the brig. If it hadn't been for Stephen we should have gone to Davy Jones.

The loss of the starboard pump proved a great handicap and I took it upon myself to see if it could not be repaired. It was an old fashioned wooden affair and when the main boom fell on it, it split off one side of the jaws, bent the pump rod and broke the handle and put it out of commission. I made an examination and found that the split did not extend more than a couple of feet below deck, so I went to Captain Soper, and told him.

I could mend it so as to make it serviceable. He gave orders for me to repair it. I got the copper to make a band of our heavy hoopiron and we put the band on tight and closed the leak in the pump, then I rebuilt the upper part of the pump with some oak timber from the hold and bolted it on to the pump, made a new handle and straightened the pump rod and in about two hours I had the pump working and throwing a good stream of water and after that we had no trouble keeping the brig free of water.

One of the first and most disagreeable jobs we had to attend to after the gale had subsided and we were able to uncover the hatch, was to get rid of the rotten blubber, which had been swashing about in the blubber room. When we took the hatch covers off- well, we all run. The blubber had rolled about in the hold for three days and nights. It was the color of dirty snow and it stunk enough to lift your head off. But we were in hot weather, it was spoiled and had to be got rid of, so we pitched it up on deck and then overboard, and then fumigated with lime and carbonic acid.

So we lost our whales, even though we had the blubber on board. We estimated that the two whales would have made about forty barrels of oil. When the foremast broke off at the keelson it smashed two casks containing twelve barrels of oil. But this oil together with that that leaked out of the blubber, probably saved the brig, or helped Stephen to save her anyway for the oil was pumped overboard and that kept the sea smooth and the waves from breaking over us and filing the deck after we were dismasted.

Saturday, September 5th. Wind still blowing hard. The brig is leaking but not badly. This afternoon we found a part of one of the planks under the stern had washed out, probably caused by the brig pounding her stern heavily on the wreckage yesterday. It was impossible to fix the leak from the outside so we were obliged to tear out the back part of the transom covering in the cabin over the break and then we stuffed part of a hair mattress into the hole.

We have partially stopped the leak, but several other planks are loose making the leak a hard one to handle until we can get to it from the outside.

Later—The wind has hauled around to the south-east and is still rough. About four p.m. we headed the brig to the north-west and shall try and make Galveston, Texas. About eleven tonight we took another big sea over the stern which washed away the binnacle and compass. The two men at the wheel nearly lost their lives, but for their being tied to the wheel frame, they might have been washed overboard. When the wind hove us down yesterday morning it was blowing from the westward, it has since hauled around to the southeast making a nasty cross sea.

Sunday, September 6th. The wind has abated considerably the last six hours, and early this morning we rigged a jurymast out of the top gallant yard, which we saved from the wreckage, and got out a new staysail and set it. We also rigged up a couple of boat masts and have the sail set on them. We are moving along about three miles an hour with a fair wind. This noon we got an observation of the sun and in the afternoon found our longitude. We are south and east of Galveston about 170 miles distant. We are trying to make that port. Brig still leaking a little.

Monday, September 7th. Has been quite pleasant all day. We are still cleaning up the wreck. A number of the crew were more or less bruised or cut during the storm and all were in need of sleep. The past twenty-four hours has given us an opportunity to catch up on the sleep we lost and rest. We have only one boat left out of the four we had before the gale. The waist boat. The wind Friday morning blowed her in on deck, davits, crains and all. She is not badly stove and can be repaired. We made one hundred

Sixteen. [Untitled]

miles towards Galveston the past twenty four hours. This afternoon we sounded and got bottom at twenty fathoms. The sea is quite green, owing to our close proximity to land. We have been heading north all day with a fair southeast wind.

Later—At 11 p.m. we sounded again and got bottom in twelve fathoms.

Tuesday September 8th. Has been hazzy and overcast all day. Early this morning we came to an anchor off Galveston Island in five fathoms of water.

Early this morning the brig Emily Watters, of St. Johns, New Brunswick, came to an anchor close to the shore. This afternoon in getting under weigh she missed stays and went ashore and at sundown was still aground but lying in a soft spot. Today we have been busy clearing the decks. The wind from the eastward hold us here as we cannot go on the wind. We shall have to wait for a fair wind or a tug.

Wednesday, September 9th. Commences with light squally winds from the southeast. This morning, at daylight, the wind being fair, we took our anchor and steered *F.H.M.* for Galveston, which is about thirty miles distant. At five p.m. we took a pilot and one hour later we came to an anchor off Galveston bar. We shall lay here until tomorrow, when a tugboat is expected to tow us up to the city. There is only twelve feet of water on the Galveston bar, but as we only draw about nine feet we shall have no trouble going in, even at low water.

Thursday, September 10th. Has been fine weather all day. Early this morning a tugboat from the city came alongside and hailed us and told us to heave short and he would come for us later. He then steamed off shore. We got ready but the boat not returning for us, in the evening we give the brig more cable and prepared to spend another night outside. This evening the wind hauled around from southeast to northeast.

Friday, September 11th. Commences fine. This morning the tugboat towed us up to the city.

We have been an object of much attention all day, many people coming off to see the first whaler ever to come to Galveston. Then our dismasted condition attracts attention also. As soon as we came alongside of the wharf we commenced to break out and cooper our oil. We also broke out water and provisions. The captain went ashore and got a quarter of fresh beef and some vegetables and we have been living high today. Tomorrow we expect to put our oil on board the "City of Waco."

Part I—"The Voyage of the *F.H. Moore*"

Saturday, September 12th. Has been a busy day on board the brig. We had all of our oil on board of the "City of Waco" by three p.m. This afternoon I secured my passage for New York on board the steamer. It will take fully six weeks to repair the brig and get her ready for whaling again, and I am tired of her, anyway, and want to go home and rest a while. So I have concluded to leave the *Moore*. This evening in company with others took a stroll over the city.

Sunday, September 13th. Commenced fine but in the afternoon it turned out rather squally. Early this morning I took my chest went aboard the "City of Waco" and started for New York. Captain Soper and his wife, Mr. Huzzey and Louis, the Portegese boatsteerer are on the steamer also. We bid goodbye to the little brig which for eighteen months has sheltered us, on which we spent so many happy days.

Monday, September 14th. Fine weather and steamer making good progress, but is was Wednesday before we arrived a Key West. Here we discharged some cargo and took aboard some passengers and some Naval prisoners, who were being taken to Brooklyn Navy Yard. At two p.m. we left for New York. We sighted Sun Spit Light and by sundown sighted Sombrero Light and at nine o'clock sighted Aligator Light. At midnight the light off Cape Florida was in sight. Thursday there were quite a number of sails in sight. Everything looks pleasant and from appearance we shall have a fine run.

Friday, September 18th. The weather was fine all day and we have been on deck most of the day watching the sails of which there were many in sight, most of them working to the southward.

Saturday, September 19th. Early this morning we passed the latitude of Cape Hatteras. One sail in sight this morning, a barkentine heading south. At one p.m. we spoke to the brig Ada Hall. of Annapolis N. S. The brig had collided with a schooner two weeks back and had lost her bowspirit, both top sails and the head of her foremast. She was working her way slowly to Philadelphia and wished to be reported as doing well

Sunday, September 20th. Commences with light foggy weather. About ten o'clock this morning we sighted Barnegut Light House about 65 miles south from New York. The ocean is dotted with sails. At four o'clock we arrived in the city and in the evening we all took cars for Boston. After an all night ride we arrived 1n Boston where I bid good bye to my shipmates and took the evening train for Portsmouth, N.H. where I arrived about nine o'clock. So ends the voyage of the brig *F.H. Moore*.

Conclusion

The *F. H. Moore* returned to port in Boston in August 1875. It went on two more recorded whaling voyages, the last from 1877 to 1879. Captain Soper led one of those voyages starting in October 1875. Almost two years later the voyage ended with only 90 barrels of oil to show for its efforts. Soper sold his house in Provincetown and moved to Melrose, Massachusetts.

Sam Williams did not linger on land either. Shortly after the end of the *F.H. Moore* voyage he shipped out on another. Again he recorded his impressions, not all of which have survived. The partial record of that voyage is in Part II.

In 1891 Sam Williams started the *Shamokawa Eagle* newspaper in Washington state. For the rest of his life he ran the newspaper and contributed to the small community on the Columbia River about twenty miles inland from the Pacific. Sam Williams died in 1925.

PART II: AN UNFINISHED BOOK
BY SAMUEL GRANT WILLIAMS

The following was written in pencil of several pieces of paper. Some of the edges are frayed, removing parts of the text.

Donnelly Sept–24th 1883

Dear Mother

I send you some of my manuscript for my book, I have copies it off and corrected it and it may interest you. Did you read the whole story in Mins [the Minnesota Star?] Tribune about Capt. Peaks you will find it the supplement of that paper. We are all quite well. Love to all.

Grant

Written on the back of the page:

Down on the quay the vessels were often stopped by Customhouse officials and actually obliged to leave any over abundance of tobacco the might have hidden about their clothes. A small allowance was given and the balance, marked, and tied in a bundle. Was held until such a time as Jack should call for it on going aboard his vessel. But is spite of this precaution Jack manages to smuggle ashore enough to get gloriously tipsy. He often [illegible] the [illegible] by spending the night in the calaboose for the Authorities are very strict about good [illegible] and considering the fact that liquors are plentiful here there is very little intoxication. One can purchase the island wines very cheaply, and the laborers drink vast quantities of them. Nearly every store has liquor for sale, but there is very little treating going on.

Fayal has a number of small stores at which one may purchase some fine goods but their owners never think it necessary to display their wares to be able to sell them. The customer always calls for what he wants and such a thing as forcing sales are not heard of here. We stopped at the store of a little man who sold fancy baskets and feather flowers and thought could his goods have been displayed, they could have had a better sale. But it owner thought different and perhaps with the island custom and his established trade, and no competition he done very well.

There is an American Consul here and a very good hotel at which all Americans meet. It is kept by a Mr. Edwards an English man and is well patronized. The harbor which is [illegible] and very dangerous, is about to be protected by a breakwater which when finished will render Jack safe.
S.G.W.

Chapter One

We have been laying in a heavy fog for two days off the island of St. Michaels Assores and have at last taken a brisk southeast breeze, and been running down the southern slope of the island at a rate that seemed to promise a dinner inside the breakwater of Porto Helgaville. We were comfortably seated at a table in the cabin of the old bark *Maret* enjoying as best we could our rather comfortable breakfast of lob-scourge [illegible] salt meat and coffee, when suddenly we felt the visible hill to starboard and heard the boatsteerer in charge of the deck hastily giving orders to take in sail. It took but a minute to clear the deck and grasp the situation. We had been struck by one of those living squalls that often come off from the land, and were turning through the water at a fearful rate, crowded over on our [illegible] by the force of the squall, and the heavy amount of sail we were carrying before the squall struck us. To take in sail on a whaleboat requires but a few minutes owing to the large crews carried by that class of vessels: but we have busted two or three light sails, and hoisted the front-fore-topsail thereby spitting that sail before the squall parted company with us. We immediately set to work to repair damages and inside of an hour had our sail set once more.

It was my duty as mate to see to the repairing of the busted top sail and as we were all very anxious to get into port I went aloft with some of the men to see to the repairing, seated on the top sail gained a scene of beauty met our eyes; between us and the land some three miles distant sparkled and flashed a sea of emerald and silver which seemed to have been crystalized by the recent squall, and in the background the Gnam [illegible] island shore resplendid under the bright sunshine; at various points different colored expanses of land shone as if lately treated with a coat of varnish, and the mountain steams refreshed by the recent squall ran madly down to the sea, where they mingled with the salt spray of [illegible] ocean.

Part II—An Unfinished Book

Our breeze refreshing we were off to the harbor of Porto Delgardo by noon and at once took our morning gang on board. The health officer and customhouse officials made a visit, leaving one of the latter to look to our cargo. They took nearly all our tobacco on shore after giving us a receipt for the same. We now proceeded to moor our vessel and soon had the good old bark fastened [illegible] and stern. The captain and cook went ashore and the latter returned with a good supply of fresh provisions; which were a great treat for we were some six months out from land. We shall lay here a month to [illegible] our vessel and ship our oil to the United States.

Chapter Two

While we are on board our bark and laying at mooring let us look around on our neighborhood. Trim English freighters are here by the score, and fine little boats they are too: all iron vessels and sharp as a wedge. Most of their topsail schooners with an occasional steamer, all laying here waiting for their cargo of oranges. They make very fast runs between ports and as one of the crew remembered "looked as if they took a dive in St. George's channel and come up off the breakwater of Porto Delgardo."

This breakwater of which I have spoken so much is worth a deal of consideration, as it is the one great feature of Porto Delgardo. Previous to its construction the harbor was open coastland and was very dangerous during southerly winds; and many and vast are the stories told by the natives of the wreaks that have ground to pieces on the highest rocks. So the people commenced to build this breakwater, this artificial harbor and when I last called there in 1878 the work had been progressing for twenty years, and although it had reached a state of absolute safety for all the vessels that might seek its shelter still the engineer in charge did not then think he could forget his flaws for twenty years to come.

The breakwater when I saw it in 1878 was about half a mile long and I should think one hundred yards wide, but the magnitude of the undertaking will be appreciated when one considers the great depth of water—from six to ten fathoms—to which they have been obliged to descend to lay the foundation. Imagine an immense wall of masonary half a mile long, one hundred yards wide and nearly one hundred feet high, and you have some idea of this vast undertaking. The amount of labor and engineering skill employed is something prodigious.

A gang of sub-marine are continually at work, and a narrow gauge railway has been constructed to run two miles into the country to transfer the stone which is used, A little steamer, the Astor, by name, flies constantly back and forth, towing heavy scows filled with rocks to be dumped in the deep water off the [illegible] of the whenever trains cannot back down. This breakwater besides affords shelter for the vessels, serves a delightful promenade to those who love a good breath of fresh ocean air, and on Sundays is visited by scores of handsomely dressed ladies and gentlemen. When it is finished I understand there will be a fine carriage drive laid out on its top, the view from which is charming. From this elevation the island of St. Mary is in full view on a clear day although distant some thirty miles.

Chapter Three

In my last we took a somewhat lengthy look at the artificial harbor of Porto Delgardo, and now let us take our first trip ashore, for we have really not been on old terra firma as yet. The vessels were packed so closely inside the breakwater that we were obliged to take our whaleboats on shore during our stay, so we engage one of the numerous boats that fly about the harbor manned by a couple of little boys, and go on shore. We will land at the new market as the water is very smooth and the boys look extremely tired, for the little fellows work very hard, and pass up by the market which had not as yet exhibited any very commercial aspect. It take a long time in the Azores to make a business popular—or to introduce new ideas—consequently the new market had not at the time of our visit received the patronage of its older competitor on the Rue de Collegeo. The old market was charmingly situated amongst an uncut grove of shade trees, which may have had something to do with its commercial success.

But to go back to this new market, it was built of white cut stone such as this island quarries, a kind of marble which presents the appearance of having been heated in some substantial furnace; the style of architecture is sometimes substantial and eloquent. It is two stories higher, the upper story bearing a resemblance to the Gothic; it is furnished with comfort and shelving for the proper display of goods, arise at the entrances. Just inside an ornamental floor, stands a pretty little office at whose open window the vendors of goods pay a rental for the use of their booths.

A brief walk of a few blocks and we come to the Lawford S.C. Francis [illegible], a very pretty breathing spot of 10 acres, covered with trees, and ornamented with foliage plants and cultivated flowers, on this island is erected a pretty bad stand and in this little home every Sunday afternoon the government band—which by the way was a very good one—drew forth a large audience of gaily dressed people. Children flocked around the flower decked banks of the lake and fed the gold and silver fish with bread crumbs.

The scene reminded me very much of Boston Common. On the south of the park stood the great stone church of St. Joseph and the hospital connected with it; on the west the only convent allowed on the islands. Then north the English hotel, a two-story stone house, noted for good dinners and fine old wines, and sporting enough bricks in two of its chimneys to furnish Stevens County with a brick schoolhouse.

East of the park stood a long street of low stone stores and orange lading houses which ran in a southerly direction past the old fortification, which never would have been taken for a fort, but for the occasional firing of a salute in honor of King Louie or some of the many Latin saints. This street terminates in the sand and of the breakwater and at this point stands the light house and the harbor masters office and a few steps from there and right at the very head of the harbor is the watering place where the vessels take on board the fresh water required to make a voyage, the whalemen sometimes taking thousands of gallons. At all times during the stay the boats carrying water could be seen at this fountain and it was here that one could study Portugesse boatmen to one's heart's content. Here too the divers at work on the lower end of the breakwater landed in their boats, and we often saw them walking around with their armor on; an odd looking spectacle.

Chapter Four

At the eastern end of the Rena de Commerca, a long street which runs parallel with the water front, stands the grand white prison, a massive two story stone building which may be seen many miles at sea towering high above the neighboring buildings. We stopped at the gates one afternoon on one of our walks and asked the turnkey if we might look over the building.

The turnkey, an old white haired gentleman, invited us in, volunteering the information that there were no prisoners in the building at that time. He stated that crime was not very present on the island, never having had but one murder during his twenty years incumbency. From the top of the prison the view was grand, the island of St. Mary, thirty miles distant was plainly to be seen; the ocean dotted with sails and boats and vessels made a very fine marine view while the view inland over the land was charming.

From our high elevation we could laugh at the blank stone walls which surround the orange orchards. These walls are often ten to fifteen feet high and in many instances their tops have a barbarous coating of broken glass bottles mixed with the mortar. Inside we find that constant years of vegetation and constant addition of fresh soil from outside has raised the surface of the gardens in some instances three feet above the surface of the ground outside.

The view in some of these orange orchards present a long line of well trimmed trees; they average about the size of our common apple tree and in their fall about November when they are loaded with fruit and very pretty. The fruit for transportation is picked quite green and is taken to the Porto Delgardo to the packing houses where a large force of girls and children wrap and pack them in the boxes for transportation. These orchards are scattered over the whole island and in the fruit season one sees a constant stream of donkeys each loaded in a way which would have provoked the ire of the Society for the P.C.A. to a very alarming extent. The Azorans think the 'burro' was made to beat and he never spares the rod, or I might say club. But to go back to the orchards, in many cases they are laid out as beautiful summer gardens, pretty streams of water are often brought through them and in a number of cases fountains have been erected and the generous owners of these gardens [the rest is missing].

[An Incomplete Chapter]

Sharing the fishing grounds with the hawks are the boats of the native fishermen. Each has its crew of ten to twelve men and boys and are always accompanied by a custom house officer whose duty it is to keep them from smuggling and deserting the island. Few Americans can really understand this latter clause, so for their information I would say that all Portugeses subjects are expected to serve in the army but a good many were in the habit of running away, coming to America and for a space of time covering that of our civil war the Portugese authorities of the islands found it difficult to furnish their quota of young men to serve King Louie.

Across the channel Pico looks down upon this scene. At one time presenting its gray rocky top, a sign of fine weather; at other it is hid for half its length, or at least that part that lies above the line of vegetation. At such a time the view from the quay of Horta in Fayal is a magnificent one, the channel between the islands, some three or four miles wide, being white with seacrests and foam, the the sun which occasionally breaks through the clouds seems to convert the mad sheets of water into a carpet of gold and silver.

Early every morning the Pico packets, loaded with wood, fruit, [illegible] comes dashing madly across over to Fayal, each boat groaning and creaking under its burden of cargo and canvas, two large [illegible] sails, half a dozen heavy outlandish oars and a crew of noisy boatmen. Arriving at Fayal the packets proceed to unload their cargo, and if it is a rough day and the boat cannot go along side of the quay, the passengers and cargo are landed at a neighboring beach, on the shoulder of the boatmen, who stagger through the surf with their heavy burdens, little caring whether it be the fat priest of one of the many little churches or a coop of cackling fowls.

We sail into the harbor of Horta and drop our anchor, taking good care to go in well towards the town. As the water is very cold and vessels that have let go their anchor too far out, have been known to lose them. This is a [part of the page is missing] northern part of the bay that has been named the 'pit'. It is said to contain a number of fine moorings. Almost as soon as our anchor touched bottom we are called to the side of the vessel by the health officer and harbor master. We are subjected to a rigid examination by those high functionaries, who place a custom house officer on board to guard against smuggling and then are allowed to go on shore.

The water being smooth we will land at the quay. Here we see the hucksters sipping their Jackass cheeses in the salt water to harden the rind. These little cheeses are made of skim milk and will weigh about a pound each. The old women find a steady sale from them as jack has forged them to be an agreeable substitute for salt meat.

Our first errand on shore will be a call at the post office, although we do not as yet expect a letter. On our way there we are beset by a half dozen beggars of every kind of hideous deformity, beggars whose impudence can hardly be equaled by Barbadoes darkies. Adroit beggars they are too. Never offering to take less than a [illegible] five cents, from a foreigner but thankfully accepting a boiled chestnut from their own countrymen. We reluctantly pay this enforced private tariff and proceed on our way. We find the post office to be a little, low one story stone building with but one window and a door, a high rail and behind it a little spectacled Portugese who presides over the mail with a Jacksonian dignity which we confess we hardly expected.

In the old whaling days, when Fayal was a rendezvous for the North Atlantic fleet what scenes have occurred in this little stone building. What happy faces to the recipients of letters from loved one at home, were wont to wear; and how well the writer remembers the disappointment he felt on his first voyage on going ashore and ransacking in vain the little office and its files of old dusty ship letters, looking in vain for the one from Mother in New England. But when a few days later the good Bark Fredonia from Boston arrived and brought the long waited for mail how gladly he received his first sea letter.

Two or three blocks further on and we come to the market, an enclosure of two or three acres, sheltered by trees and booths under which the market women display their various wares. All kinds of vegetables, the finest among them the finest being the great red sweet potatoes and silver skinned onions. All kinds of fruits, oranges, apples, figs, grapes, pineapples, and chestnuts as large as our American horse chestnut. A large variety of earthen

[An Incomplete Chapter]

dishes and wares and the noisy wooden shoes so much wore by the farmers.

We rest awhile under the shade of the trees and eat a quantity of grapes and green figs. We watch the people making their bargains. Some of the grooks are very grotesque and with their musical tone of speech they render the market scene a pleasant one. We remember on one occasion going on shore with the mate of the vessel. He was a Portugese but liked nothing better than to play a joke on his countrymen. he walked over to the market and sitting down amongst the baskets of fruit commenced to eat, the hucksters encouraging us for they know we are good pay. After we had eaten all we can the mate proffers an old woman, the owner of a big basket of grapes, an American 5 [cent] piece, a nickel one. She examines it curiously, bites it, turns it over, tries to look through it, endeavors to bend it and finally taking it to another old woman the two begin an animated discussion as to its merits and value.

Very soon they are joined by others, and soon a crowd gathers. After a while the offending piece of money is carried to a policeman, who after a very minute inspection concludes that it is American and not the class of money passable in the realms of King Louie. The "In God We Trust" must be especially American for what good Portugese ever saw an inscription of that kind on the money of the Azores? He hesitates a moment, and then with a strong sense of unpleasant duty to perform steps boldly over to where us offenders are sitting followed by nearly every old woman in the market. "Do you know that we are in the land of King Louie of Portugal?" and "Are we aware that we are passing base money on the subjects of the King?"

It is beginning to become anything but a joke now and our mate in the excitement his to unmask and explain, which he does much to the delight of the old women, who upon receiving the pay for their fruit in real coin of the realm, turn upon the policeman and upbraid him for having insulted the distinguished strangers. It is needless to say that the guardian of the law escapes as soon as he possibly can from the terrible tongue lashing of the ignorant hucksters.

Quite a number of little villages are scattered over the island, not the least amongst them the little harbor of Port-of-Pines. It is a very good boat harbor and occasionally one sees a small vessel anchored there, but in rough water it is dangerous. There are some old fortifications here which in olden times may have been considered impregnable. That was in the days when the bold corsair sailed the ocean blue and often went so far as to demand a ransom from small seaports. We were told that during the war of the rebellion an American schooner was confined in this little harbor by the privateer Alabama, and was obliged to stay in this strip of neutral water for a week or more.

The arrival of a terrible gale of wind drove the pirate off to find shelter under the lee of St. George, leaving his victim to be dashed upon the rocks. The little American, however, had no intention of surrendering to the elements and as soon as the pirate was out of sight he succeeded by hard work and pluck in escaping, but only to be caught a few days later by the pirate who out of admiration for the captain of the schooner it is said allowed him to go on his way unmolested.

Wandering around over the island one meets with a little stunted evergreen called "fayo" from which the island is said to take its name. Hedges of horteucia with its flowers of white and purple line the roadside. Tall stacks of corn drying in the sun and pretty gardens of cultivated flowers. Little trickling streams in whose waters the native women were busy washing. pounding their clothes with a heavy wooden paddle, which must have been hard on the buttons if nothing more. We stopped a moment to watch the women at their tasks and was surprised to be greeted with the following question delivered in recognizable English "American women no likee wash?" The young woman stated that she had "been to America," and had worked at a cotton mill in Lowell, Mass but that she liked the old home at Fayal better than the land of the stricken [illegible]. She had been a willing bread winner but had found her associates in the mill too rough for her. We have been listening for some time to a terrible squawking, and as it draws near we recognize the ox cart of the Azores. Imagine the most primitive of vehicles, a rough framework balanced on two wooden sticks, a body resembling a crockery crate, a pair of thills with an ox to propel the mass and you have a typical Azoran conveyance. But hold: I have forgotten the squawk, no good Azoran ever thinks of greasing his cart and the result is a screech that would make the dying lament of a pig seem like the music of an Eolian harp in comparison. We have been told that some of the Jelues boast of a squeak familiar to their carts also, that the Azoran has no nerves and we honestly believe it.

On arriving back to Horta after our stroll we hear the Angelus bells ringing and meet the devotees going to the churches. Some of them bare footed and carrying their shoes in their hands to be donned at the entrance of the church. The different styles of dress are interesting to study. They range from the most primitive home made island costume to the most elegant made in Paris. Some of the women wear what is called the Capota, it resembles the outlandish, overgrown dense forest. It is made of whalebone and heavy blue cloth with a long cloak to match, which covers the white dress. This gives the woman a very man like appearance. It must be far from comfortable on a hot day.

Canary Islands

In changing the scene of my journey from the Azores to the Canary islands, I shall also change vessels. I shall figuratively speaking step from the deck of the old bark *Ware* to that of the schooner *Wm. A. Grozier*. Although the voyage in the schooner was made some three years prior to that of the Ware.

I find by reference to my journal kept in the schooner that we were off Fuenteventura the last of October, 1875. The first land we sighted was the island of Lauzasota, which was north-northwest of Fuenteventura. It looked bleak and rugged from our point of view, some thirty miles distant, [illegible] safeguard inquiry has confirmed that information.

We next sighted Fuenteventura the English translation of which is "Venture of Forces." The island is some forty miles long and twenty wide. There is a revolving light on the west side which may be seen at sea some ten miles. The coast is rocky and wild.

[Illegible] Island there is a town of some 2,000 inhabitants. They claim to be of Spanish descent. We were told that back from the coast, about five miles, there were some very fertile valleys and that the people acquired their living from agriculture. Lying at anchor off the storehouses, on the [part of the page is missing] the land looked as if it would be a hard job for the goats of which we saw with a number of flocks, to get a living. The surf breaking against the rocks with a heavy roar, but as heard from the vessel it sounded not unpleasant. We saw on approaching the beach off the storehouses a brig lying at anchor boiling out a whale. It turned to be the Brig D.A. Small of Provincetown. The next day we saw a large whale and we lowered and chased him all day. In the afternoon the Port boat went on, but did not succeed in fastening to him.

The next day we cruised off and on the land and the following day came to an anchor near the D.A. Small. We spent a day or two painting the vessel and repairing our sails. On Sunday the Captain let us take a run ashore to see the place. We landed at the storehouses on a strip of hot white sand. It must have been 120 [degrees] in the sun, there was no shade and we walked some fifty rods or more up to a cluster of perhaps a dozen low stone houses. They were covered with the usual coating of lime and it almost scorched ones eyes to look at them for any length of time. On nearing the houses we found a lot of [illegible] natives engaged loading a herd of several camels whose plaintive cry betokened their displeasure at being overloaded. Some of the camels were standing in the sand with their knees strapped together—a way the Arabs have of fastening their stud while on the desert. It is said that the camels will not move if overloaded and that they will almost suffer death rather than journey under such circumstances. In this instance we thought they were carrying more than they should, but perhaps they knew their journey was a short one. They are very fond of the prickly pear and as far as we could see there is nothing else for them to eat.

We noticed that quite a number of the houses were devoid of floors, the white sand of which they were built serving that purpose. A saloon with its dangerous contents was here, and seemed to be well patronized. They make a liquor here called Anisese, [Anise seeds] being one of the principal ingredients. For general effect it seemed to answer all the purposes of a more expensive liquor. All kinds of provisions were had at very high prices, so we done no trading here. Teniriff is forty miles from Fuenteventure and we made the passage by night. It was a beautiful night; bright starlight succeeding a short and brilliant twilight, the sun going down in one of the most beautiful sunsets we ever saw. A perfect zone of gorgeous clouds haunting the whole western sky.

We were [illegible] on our course by light trade winds and on our vessels bow slowly drift the clear, blue water, the surroundings were such as might have induced a dreamer to think of fairy-land. Overhead the dome of the twinkling stars shone brightly. The constellation of Ursa Major making its circle around Polaris, Cephas and Capellia, Series the dog star and about eleven o'clock the "bright reagent of the heavens" that Cafe Cochoran appeals to his dilema rose almost full and sent a silver glaze over the ocean. [Part of the page missing but probably It] was a night of ocean beauty, and when at twelve o'clock the man at the wheel struck eight bells and we had been relieved we were told to leave the deck for the rather close quarters of our room.

We came on deck again at four o'clock just as the daylight was beginning to break. Only the brightest of the planets were visible and the moon was fast losing her brilliancy. The Trade winds were freshening and light clouds were chasing one another towards Tenerife whose top was already hidden. The island of Gran Canary lengthened out to the south of us, until it presented vineyards and farms. At length we arrive at the harbor of Santa Cruz. We pull into the harbor passing by a great French corvette. We go in and find we will not be allowed to land unless we come to anchor. So we only leave our letters and bidding the American consul's clerk—who kindly consented to mail them for us, good by; go on board the schooner and hoist away for the coast of Africa. There were a number of whaling vessels laying at anchor.

Editor's Note

Whether Sam Williams never finished the manuscript or if the rest of it was lost is unknown.

Part III: Excerpt from *Incidents of a Whaling Voyage* by Francis Allyn Olmsted

Francis Allyn Olmsted was a student in 1839, recovering from an illness. To rebuild his health and to escape a harsh New England winter he signed on as a passenger on a whaling ship for a voyage to Hawaii and other ports in the Pacific. In this first chapter of his 1841 book Incident of a Whaling Voyage, *Olmsted observes the operation of the whaling ship* North America. *The asterisks (****) indicate an omission from the original text.*

Friday, Oct 11, 1839. Early this morning, the rattling of blocks and rigging, and the animating cries of the seamen, announced that the North America was getting under way; and soon the barque with her swelling sails distended by a gentle breeze, swung from her moorings. The wind was fair, and as we glided out of the beautiful harbor of New London, the clear air of the morning, the favoring breeze, and the bright sun mirrored in a thousand tiny waves, soon dispelled the gloom of parting from those I loved, and even inspired me with renovated spirits. The band of the Revenue Cutter was going through its morning exercises, and I listened to the national airs it was performing, until growing fainter and fainter, they were lost in the distance.

A new feeling of patriotism was awakened within me; and these simple strains, that on ordinary occasions, would scarcely have been heeded, were now associated with many endearing recollections, and invested with a melody and sentiment I had never before discerned in them. Month after month will perhaps have rolled over me, ere I shall again hear the inspiring strains of "Hail Columbia, happy land" in my own favored country to which I am now bidding adieu—it may be forever. But from these painful suggestions that now and then struggled to obtain possession of my mind, I turned with interest to the scenes as they opened before me in my new habitation, the first aspect of which was not the most favorable.

The North America is a *Temperance* ship; that is, no ardent spirits are served out to the men on any occasion. This, however, does not preclude them from becoming intoxicated whenever an opportunity presents itself, which two or three of them, judging from appearances, would not be very

Part III—Excerpts from *Incidents of a Whaling Voyage*

reluctant to embrace. The prospect of a voyage of three or four years in length is an incentive to greater excess, while intoxicating liquors can be purchased to drown the unpleasant anticipations incident to so long a separation from country and kindred.

Inebriety is by no means as prevalent among sea-faring people as was formerly the case, since the abandonment of the idea that intoxicating drinks were indispensable to the sailor. It has been within a few years only that the plan of sailing ships upon temperance principles, has come into extensive use; before this, if a master of a ship, in visiting another, declined a glass of spirits, his refusal was regarded as an insult. Soon after the commencement of the temperance reform, Major Williams, of New London, determined to lend the weight of his extensive influence in promoting temperance aboard the whale-ships sailing out of this port, in which he was interested. His exertions, although meeting with great opposition at first, were successful—other influential men followed his example, and now, out of the thirty or forty whaling vessels belonging to the port of New London, almost all are navigated upon temperance principles. To the credit of the American Whale Fishery, it ought to be added, that the proportion of vessels of this character, is much greater in this service than in any other department of our marine.

This afternoon, as I was standing at the starboard gangway, watching the progress of the ship through the water, a sailor passed by me, and letting himself down the side of the ship by the chains, very deliberately threw himself overboard, and commenced swimming towards land.

****

The wind, which during the day, hardly moved the ship through the water, as evening came on, veered ahead. A head tide also, opposed our progress, and as the sky towards the south-east looked lowering, with some indications of a gale, it was thought advisable to return. The ship's head was soon pointing towards New London, distant about twelve miles, and we came to anchor two or three miles from the shore, where we lay during the night. Early on Saturday morning, as the wind continued to increase from the south-east, we hauled in opposite the light-house.

Sunday, Oct 13. Soon after the ship was moored, yesterday, I went ashore with Captain Richards and the pilot, where we remained until this morning, when at an early hour we were summoned on board ship, as the weather seemed favorable for going to sea. But our expectations are disap-

Part III—Excerpts from Incidents of a Whaling Voyage

pointed, and here we lie without breeze enough to carry us out, while a damp atmosphere and cloudy sky, render our situation extremely dismal. It is the Sabbath too, and while the solemn tones of the distant church-bell should awaken emotions befitting the day, our own unpleasant situation engrosses all our attention; and instead of occupying our minds with the solemn duties of the Sabbath, we are watching the clouds for indications of fair weather.

Monday, Oct. 14. "Boat-ahoy," hailed the officer of the deck, as a boat was seen coming down to us, rowed by two boys, carrying a large bag in the bow of their tiny craft, intended for the ship. We were endeavoring to divine the contents of it, which were supposed to be of a highly valuable character, from the important air exhibited by the boys. The bag was hoisted upon deck and opened, when out jumped an old cat and her numerous progeny, that ran squalling around the deck to our surprise and diversion.

Cats are consequential personages onboard, as they protect us from the depredations of huge cock-roaches that swarm in every direction. I found one of these erratic black-legs the other day, up in the main-top, wandering about very much at his leisure. Capt. R., a few days ago, in speaking of the good qualities of the *North America*, said that "she was built entirely of live oak," which subsequent observations have fully verified!

Last evening, the clouds for a short time dispersed, and the stars and the moon beaming forth, seemed to promise a favorable change in the weather. Not long after, however, the sky was again overcast, and before morning, an easterly storm came pattering down upon deck, with the gloomy prospect of another dismal day. If I had not started with a good resolution to be disconcerted by nothing that might happen, I should by this time have been tempted to give up an enterprise so inauspiciously begun. "So much for sailing on Friday," an old salt would say. There has been a singular superstition prevalent among seamen about sailing on Friday; and in former times, to sail on this day, would have been regarded as a violation of the mysterious character of the day, which would be visited with disaster upon the offender. Even now it is not entirely abandoned; and if a voyage, commenced on Friday, happens to be unfortunate, all the ill-luck of the voyage is ascribed to having sailed on this day. An intelligent ship-master told me, that although he had no faith in this superstition, yet so firmly were sailors formerly impressed with superstitious notions, respecting this day, that until within a few years, he should never have ven-

Part III—Excerpts from *Incidents of a Whaling Voyage*

tured to sail on Friday, for the men would be appalled by dangers which they would think lightly of on common occasions, and their efforts would be paralyzed by their imaginary fears of being under a mysterious and malignant influence. I have been told, that several years ago, a ship was built and sent to sea, to test this superstition, and convince the craft of its folly. The keel of the ship was laid on Friday; on Friday her masts were set; she was completed on Friday, and launched on this day. Her name was "Friday," and she was sent to sea on Friday; but unfortunately for the success of the experiment, was never heard of more.

As knowledge advances, all opinions not consonant with reason must be abandoned, and this superstition is fast losing its hold on the minds of sea-faring men, especially since the establishment of the packet lines, and the frequent necessity of sailing on Friday. It had its origin, I am told, in the ancient custom of executing criminals upon this day, which imparted to it an unlucky character. I have also heard it ascribed to a connection with some of the observances of the Roman Catholic Church, which entertains some peculiar notions with regard to this day.

Tuesday, Oct. 15. Rain—rain—rain—with a raw wind from the northeast—cold and cheerless on deck—damp and dismal in the cabin. For our encouragement, the barometer, which for the last three days has been continually falling, is now rising, indicative of fair weather.

This morning, hearing an unusual noise upon deck, I ran up the companion way, and, at the distance of thirty or forty yards from the ship, saw one of the men making desperate efforts to reach the shore by swimming. One of the boats had just been lowered—pursuit was instantly made, and the man with but little resistance was secured and brought on board, crestfallen enough, in his dripping clothes, with his shoes tied around his neck.

"Come here," said the commanding officer, (the second mate) in an authoritative tone. "Well, you were going to leave us in the lurch, were you?"

"Why sir, Mr. L—— (the first mate, who was on shore) told me I might go ashore with him, and he went off without me."

"And so you thought you'd work to windward of us in this way, eh?"

"Why sir, I thought he didn't do what was right."

"You thought? Well, I'll tell you what *I* think, and I'll inform you in the most delicate manner, that if you show any more of such fandangos here, you'll be clapped down into the lower hold, sir, with some irons around your wrists, that don't look quite so pretty as ladies' bracelets neither—bear that in mind, and be off, sir."

Part III—Excerpts from Incidents of a Whaling Voyage

The crew, though very quiet in general, are beginning to show signs of impatience, and if there are no indications of fair weather at sunset, an attempt will undoubtedly be made to desert during the night. With the few exceptions I mentioned before, they are very temperate, and I have heard but little bad language or profanity on board, both of which are prohibited by the Captain.

Capt. R. left us last Sunday evening, and has not yet returned. I should have accompanied him up to town, were it not that I had already bidden my friends "good-bye" three times, and did not like to impair the virtue of the "Farewell" by repetition.

Wednesday, Oct, 16. Yesterday afternoon the clouds began to break away, and the sun shone forth to glad den us after a long absence of his cheering beams. The moon, too, favored us last evening with her kindly radiance, and long I paced the deck, musing on the reality of the enterprise in which. I had embarked. When we are preparing for a long voyage, we talk of separation from home, kindred, and country with a kind of vagueness as if it would never be realized; but when we have actually embarked, and there is no return, then the reality comes vividly to mind, and impresses us with the magnitude of the enterprise; while the uncertainties of the future forbid our anticipating its termination. The future to me is more than ordinarily uncertain. To picture to myself my various wanderings over the mighty ocean, in accommodating myself to the erratic life I have now chosen, and after leaving my present shipmates to trace out my circuitous course back to my native land, is beyond the reach of mortal ken and were a vain attempt. And there are solemn musings too. Ere I return, the irrevocable hand of death may invade the home of my youth and the circle of kindred friends, and consign one or more to the grave! Ah! these are the saddest thoughts, that press like an incubus upon the spirits of the voyager as he leaves his native shores.

****

Thursday, Oct. 17th. The wind has been light and baffling since yesterday. This noon there was a perfect calm, and upon the eighth day from the date of our first setting sail from New London, we find ourselves at Anchor off Montauk point, to prevent being drifted ashore, instead of tossing about upon the Atlantic one third of the way across.

All hands have been engaged in various duties about the ship, such as overhauling the spare canvass, and stowing away articles more compactly.

Part III—Excerpts from *Incidents of a Whaling Voyage*

The boats too, have been put in complete order, to be in readiness for the first opportunity that presents itself for using them, and although it may be a deviation from the plan I have adopted, I cannot do better, perhaps, than to describe the whaleboat and its various appurtenances.

The *whaleboat* is a narrow, light built boat of about twenty-five feet in length, sharp at both ends, with its sides gracefully curved and running up to a point fore and aft, and from its construction, is expressly adapted to great velocity of motion and safety among the swelling billows of the ocean. Unlike most ship's boats, it is clinker built, as this peculiar mode of construction is called, i.e. the thin boards that cover the ribs overlap one another, thus giving strength to the boat and enabling it to be made much lighter. Each boat is fitted with six oars of various lengths. The steering oar, usually from twenty to twenty two feet long, is confined to the boat by a strap passing around it and attached to the sternpost. This gives the helmsman great power over the movement of the boat far superior to the steering with a rudder.

The thole pins, between which the oars are plied, are covered with matting, so as to prevent any noise in the motion of the oars. Of the offensive weapons, the *harpoon* is the most important.

The harpoon is an iron instrument, about four feet in length, terminated at one end, in a sharp barbed head, and at the other, in a socket for receiving the "iron pole," a heavy wooden handle of about equal length, which gives to the instrument great momentum. A strap with a turn around the socket of the iron secures it upon the pole. To the strap is attached the *line*, a strong rope about two hundred fathoms long, which is carefully coiled up in a tub placed in the afterpart of the boat; and going around the "loggerhead," a strong post projecting above the stern, passes through a "*chock*" or grove in the bow of the boat, and is "bent on" to the harpoon. Each boat usually carries four or five harpoons, two of which are always ready for immediate use when the boat is in pursuit of whales. Their barbed heads lie across the bow of the boat, with their shafts resting upon two "crotches," or spurs, standing out from a stick rising from the side of the boat. This position gives steadiness to the weapon, and it is close at hand whenever opportunity offers for using it.

The *lance* is two or three feet longer than the harpoon. Its head is of an oval shape, pointed with steel, and its shaft is long and slender, with the "*warp*," a small line about eight fathoms long, attached to the extremity of it.

Part III—Excerpts from Incidents of a Whaling Voyage

The *spade* is a short instrument, with a thin, wide blade set upon a light shaft of five or six feet in length.

These instruments are ground to a very keen edge, and kept constantly bright. Their sharp heads are enclosed in sheaths, to defend them from injury, as also to prevent their doing any mischief. A hatchet, a couple of knives, a water-keg, a lantern, and a boat compass, together with one or more buckets, complete the equipment of a boat.

Six men constitute a boat's complement. Of these, the captain or one of his mates is one, who directs the attack upon the whale. There is also a subordinate officer called *boat-steerer*, who performs the duties of a cockswain, taking care of the boat with its appurtenances. To each man is assigned an oar and a station in the boat, to avoid any confusion when starting in pursuit of a whale.

In attacking the whale, the captain or one of his officers takes the steering oar, and directs the boat in the onset. The boatsteerer pulls the short oar in the bow of the boat, and at a signal or command from the officer, draws in his oar, and taking his stand firmly in the bow, the word is given, darts the harpoon with all his strength into the whale. Sometimes he is so successful as to fix both irons, which generally ensures the capture of the struggling monster. He now exchanges places with the officer, and takes the steering oar, while the latter comes forward to thrust the lance into the vitals of the whale whenever he comes up to blow, a feat requiring no ordinary dexterity. The moment the whale begins to slacken the line to which he is "fast," it is hauled in, and coiled up carefully in the tub, while the boat is drawn towards the whale, as he comes on top of water, when he receives several thrusts of the lance in succession, which often enters to the depth of several feet. When the animal is very violent in his movements, a few strokes of the spade across the sinews of his flukes, disable these his most powerful weapon of defence and motion. The line is confined to the grove in the bow of the boat by a wooden peg, which breaks in case the line becomes entangled, thus averting the extreme danger of being instantly carried down.

Thus much for the description of the whale-boat at present, which in grace and velocity of motion, is not excelled by any ship's boat.

On board of all vessels, the men are separated into two divisions, called the *larboard* and *starboard watches*. The first and third mates command the larboard watch, and the first and third mates command the larboard watch, and the second mate commands the starboard watch. This morning,

Part III—Excerpts from *Incidents of a Whaling Voyage*

the crew were all summoned upon the quarter deck, and the first and second mate selected alternately, the members of their respective watches. The Captain and each of the officers, in a similar manner, in the order of rank, then made choice of the required number for the boat he commanded.

Friday, Oct. 18. Last evening the ship was again under way, and at sunrise this morning, land was no where visible. There was scarcely breeze enough to steady the ship, while as far as the eye could reach, not an object presented itself to break the monotony of the ocean with its ceaseless undulations, or to impair the emotions of sublimity with which vastness of extent impressed me, as I scanned with eager eye, the uninterrupted curve of the horizon. The open ocean is rarely calm, such as we see in the waters of our lakes and rivers. Even in its stillest moments, when not a breath of air agitates it, its surface is perpetually heaving as if with some internal commotion. For the fathomless waters of the ocean acquire such a momentum when the storm comes over their depths, that even when the winds are hushed, they do not soon subside.

Tuesday, Nov. 5. In resuming the thread of my narrative, which has been interrupted for more than two weeks, I cannot do better perhaps than to commence from my last date, and endeavor to give a slight sketch of what has befallen me in the meantime.

On Saturday, Oct. 19, towards evening, the rain began to fall in frequent showers from the South. About 11 o'clock that night, I was roused from my slumbers by the rolling of boxes in the cabin, and the crash of the steward's crockery in the pantry, the howling of the wind and the loud tone of command from the officer on deck. "Tumble aft! Tumble aft here every one of you. Let go your top-gallant halliards fore and aft—clew up—mind your helm—keep her off before it—main-tack and sheet let go—clew him up, clew him up—jump, for your lives, men! Top-sail halliards let go—one of you give 'em a call there in the forecastle and steerage." "All hands a-hoy," just heard above the roar of the winds, summoned the larboard watch on deck, as we sprung up the companion-way to ascertain the cause of the sudden alarm. We had been moving along under easy sail, when upon nearing the gulf stream, a heavy squall struck us from the west. The top-gallant sails and top-sails had been settled down, while the main course was flapping about with a noise like thunder.

In a short time, however, all the sails were snugly furled, with the exception of a close-reefed main topsail and foresail, under which we drove before the gale that pursued us across the gulf stream. The next day (Sun-

Part III—Excerpts from Incidents of a Whaling Voyage

day) a sea struck our larboard quarter boat, and dashed her to pieces,— a bad omen for the commencement of the voyage. We have since had another boat stove by the violence of the sea, which dashes in very frequently across the waist of the ship.

I had brought a thermometer with me for the particular purpose of ascertaining the temperature of the water in the gulf stream; but the violence of the sea put an end to all philosophical speculations. I was informed, however, by those that were drenched by the spray, that the water was very warm.

The air, too, was mild, unlike the storms we have at home in the month of October, in this respect. Indeed, the temperature of the ocean air off soundings, is always much higher than that of the land in the same latitudes, out of the tropics in the cool season of the year. For the three weeks, during which we have been at sea, we have had no weather cold enough for an overcoat, except at night, although at home, I presume, anthracite fires are glowing to repel the first approaches of winter.

In a day or two we had crossed the gulf stream, and were promising ourselves a delightful run to the Azores, when the wind came around ahead from the eastward, where it continued for eleven days without alteration. At one time we ran down as far as the Bermudas, and were admonished to alter our course by the frequent squalls that assailed us.

During the stormy weather in the gulf stream, I confined myself to my berth, as the most comfortable place I could find, and with bundles on each side of me, endeavored to keep myself from rolling about. The motion of the vessel, and the intolerable smell of bilge water which came steaming up from the hold through the crevices in my state room, brought on a disease, that for more than two weeks, completely disabled me. It was not *seasickness* under which I labored, but an extreme debility accompanied with fever. There can be no mistaking the former, and I considered myself well versed in it from an intimate acquaintance during several coasting voyages.

A determination to rise superior to my physical weakness, was the only thing that enabled me to counteract the extreme depression that assailed me, and I have never been more convinced of the truth of a saying which has almost become a proverb—that "a resolute spirit has greater efficacy in combatting our bodily ills, than medical prescriptions." No disrespect to the profession, however.

When we are sick on shore, we obtain good medical advice, kind attention, quiet rest, and a well ventilated room. The invalid at sea, can command but very few of these alleviations to his sufferings. The attentions he receives, have none of that soothing influence, which woman's tender sympathy alone can impart. Undisturbed repose is out of the question, where everything is in motion and the bulkheads are dismally creaking. The air of the cabin of a ship is always close and uncomfortable in bad weather. Let a man be sick any where else but on shipboard.

For the last three or four days, the wind has hauled us, around to the west and northwest, with frequent squalls. Hardly a day passes, but the wind comes whistling down upon us, and lashing us awhile in its fury, leaves us, to be soon succeeded by another, when the same scenes of "letting go the halliards—clewing up and clewing down —" are enacted over and over again. During the intervals, the ship rolls heavily in the sea, and the deck is washed by the sea breaking in across her waist. Buckets, pieces of wood, and other loose articles run around the deck in wild disorder, to the serious annoyance and hazard of one's nether limbs. Shower baths provided gratis for those who are not on the look-out for themselves. We have seen no whales as yet, and even if we had, the sea has been too high for a boat to venture out in pursuit.

During the frequent squalls of the few days past, I have been delighted with the beautiful rainbows that formed at *all* hours of the day—now spanning the heavens in a regular arch, then rising above the sea like two pillars of resplendent colors, and again but just tingeing the clouds with their brilliant hues.

We are now about eighteen hundred miles from the United States, and expect to reach the Western Islands in six or eight days.

Part IV: Excerpts from *Etchings of a Whaling Cruise* by J. Ross Browne

J. Ross Browne was a native of Kentucky hungry to see the world. His journey took him east where he worked for a time as a clerk in Washington, D.C. Still unfulfilled, he and a traveling companion set out to explore the earth by going to sea. After signing on to the whaling ship Styx, *a ship named after the river of hell from Greek mythology, Browne soon discovered he was on his own journey in hell. In vivid detail he recorded the hardships and misery of whalemen. When he returned to the United States he published in 1846* Etchings of a Whaling Cruise, *both an expose and a diatribe against the abuses of the industry. As Browne wrote:*

My design is simply to present to the public a faithful delineation of the life of a whaleman. In doing this, I deem it necessary that I should aim rather at the truth itself than at mere polish of style. A due regard to fidelity induces me to present the incidents and facts very nearly in. their original rude garb. I have no faith in softening or polishing stern realities. Let them go before the world with all the force of truthfulness; and if they can effect nothing, the blame will not rest upon the narrator. I claim no higher credit than that of being an accurate reporter of passing events, with the privilege of commending what is right, and dissenting from what is wrong and no fear of exciting enmity shall deter me from exercising it.

Anyone tempted to romanticize about the adventure of whaling should read Browne's full account of his voyage, portions of which are excerpted here. The asterisks (****) indicate an omission from the original text.

I well remember the night previous to our departure. It was that of the 4th of July. After the usual ceremonies of the day, there was a grand exhibition of fire-works in the President's garden. A large concourse of citizens, visitors, members of Congress, and diplomatic characters, had assembled on the terrace of the Capitol to witness the brilliant and imposing scene. Some kind friend had circulated a report that we had received a commission from his excellency, Mr. Tyler, to arrange a matter of great national

Part IV—Excerpts from *Etchings of a Whaling Cruise*

importance with the government of Portugal. The consequence was, that several of our distant acquaintances, who had formerly recognized us with a stiff nod, now crowded around us, and bid us good-by in the kindest manner imaginable, wishing us a most cordial reception at the court of Donna Maria.

Having procured passports at the State Department, we took our departure in the cars early on the morning of the 5th of July, 1842. As it was not probable we could find a vessel in Baltimore bound for Europe immediately, we continued on to Philadelphia, where we spent a few days, and obtained some letters of introduction from a friend in the Customhouse to distinguished gentlemen in different parts of Europe. Finding no encouragement in Philadelphia for tourists with slender means, we proceeded to New York.

Our joint purse on leaving Washington amounted to about forty dollars. Of course, we could not deny ourselves the gratification of visiting the various places of public amusement; besides, being gentlemen up to that time, it was indispensable that we should patronize the best hotel, ride in an omnibus or hack whenever we did not feel disposed to walk, and be liberal with servants and porters. At the expiration of a few days, it alarmed us to find that we had but eight dollars left.

Upon application for temporary employment, with a view to replenish our means, we learned that business was very dull, and young men were glad to avail themselves of the privilege of passing their time usefully in mercantile houses without remuneration; a species of amusement not particularly adapted to our circumstances. With due humiliation, let it be told, we were soon reduced to the necessity either of writing to our friends for a remittance, or of being insulted with an invitation to depend upon the charity of casual acquaintances. The first was out of the question; it would destroy our diplomatic reputation; the last was too galling to our pride to be entertained for a moment.

In this dilemma we strolled down to the shipping, and went on board a vessel bound for Bremen. The captain, a jolly-looking Dutchman, sat upon the companion way smoking his pipe, while he kept his eye upon some of the crew who were at work on the main deck. He received us very kindly, and gave us much information on the subject of seafaring life. It would be a difficult matter, he said, for two young men dressed as we were to procure employment on board a merchantman as light hands; but if we put off our "long togs," and went to work in a corn-field for about three

Part IV—Excerpts from Etchings of a Whaling Cruise

months, to give us a hardy look, we might succeed. Where there were upward of four thousand seamen idling about the wharves, it would be no easy matter for "green landsmen" to make a voyage. On the whole, he gave us rather an unfavorable idea of the life of a sailor, and advised us to try something else. He thought it a pity that young gentlemen of education should waste their time in a pursuit so little adapted to their physical strength. There were rough fellows enough in the world who could do that sort of work better than persons who had been delicately raised.

The words of the kind-hearted old skipper made a deep impression upon our minds, and, if it were not for sheer shame, and the pressing nature of our circumstances, we would have abandoned our romantic notions at once. However, we felt that we were in for it, and it would not do to back out. Nothing, therefore, remained for us but the prospect of getting something to do on board a ship. It made no material difference to us in what capacity we went; all we desired then was to take leave of New York.

The rest of that day and part of the next we spent in making inquiries at the ship agencies along the wharves; but our appearance, combined with our anxiety to become sailors, excited suspicion, and the answers were so unsatisfactory that we began to despond. I noticed that the old tars, who were lounging in groups about these offices, smoking their pipes, and chatting in a nautical style of language totally incomprehensible to us, eyed us slyly, and winked at each other as we passed. In the course of a few months we very well understood what they meant.

There was something of novelty in being thrown upon our own resources in a large city, without a single friend to whom we could look for aid. Still, as our money was spun out to a few dollars, it became necessary to leave off romancing, and bring our ideas down to the level of our circumstances.

As we strolled along one of the wharves, casting wistful glances at the vessels close by, and now and then taking a peep into the shipping-offices, our attention was attracted by a slip of paper over a door bearing the following important intelligence:

"WANTED IMMEDIATELY!!!"
"Six able-bodied landsmen, to go on a whaling voyage from New Bedford. Apply up stairs before 5 o'clock p.m."

This was somewhat encouraging. Indeed, we thought it peculiarly lucky. It suited us exactly. We stopped and read the words over half a dozen times, in order to satisfy ourselves that we were not mistaken as to their import. But here was the difficulty: the notice said *able*-bodied landsmen. Were we of that description? We consulted the matter for some time, and at last came to the conclusion that light-bodied, active men, with a considerable share of spunk, ought to succeed as well as heavy-built men. We accordingly entered the office with a bold, independent air, as much as to say, we knew what we were about. An excessively polite old gentleman of prepossessing appearance received us with every manifestation of cordiality. In answer to our inquiries concerning his notice, he replied:

"Yes, gentlemen, I want a few more men. Do you think of shipping?"

"Why, yes, we have some notion of it."

"The very best thing you can do; sorry you are not a little stouter; but no matter, I think you'll answer the purpose. I just received a letter this morning from Mr. ——, the whaling agent in New Bedford, requesting me to send on two light, handsome fellows. He don't care so much about their weight, if they're good-looking; wants them for a small vessel, you see, and likes to have a nice crew."

"Well, you think we'll do?"

"Oh! no doubt about it. I'm willing to risk you, though I may lose something by it. Whaling, gentlemen, is tolerably hard at first, but it's the finest business in the world for enterprising young men. If you are *determined* to take a voyage, I'll put you in the way of shipping in a most elegant vessel, well fitted: that's the great thing, well fitted. Vigilance and activity will insure you rapid promotion. I haven't the least doubt but you'll come home boatsteerers. I sent off six college students a few days ago, and a poor fellow who had been flogged away from home by a vicious wife. A whaler, gentlemen," continued the agent, rising in eloquence, "a whaler is a place of refuge for the distressed and persecuted, a school for the dissipated, an asylum for the needy! There's nothing like it. You can see the world; you can see something of life!"

The enthusiastic advocate of whalers then handed us a paper, which we immediately signed without reading, not wishing to give him time even to reflect upon his bargain. Promising to be at the office by half past four, we took leave of our worthy friend, and warmly congratulated each other upon having accidentally met with this benevolent old gentleman, who not only smiled upon the indiscretions of youth, but forwarded all our plans, and seemed ready to oblige us in every way. From a man whom we had never seen before, all this was certainly very gratifying.

Part IV—*Excerpts from* Etchings of a Whaling Cruise

At five o'clock on the same evening we took a passage in the Cleopatra for Providence. In order that particular attention might be paid to our comfort—as we supposed, but in reality to prevent our escape—we were consigned to an officer on board the boat. The agent, also, to enhance our enjoyment, sent with us a couple of entertaining fellows, rather rough to be sure, and not very respectable in their appearance, bound on the same delightful mission. For all this we felt exceedingly grateful to our benevolent and venerable friend. It is true, we discovered after we got to sea that he had forwarded a bill of ten dollars to the New Bedford fitter, to be placed on our account with the owners. As we had sold one of our trunks, and some other unnecessary articles, the proceeds of which enabled us to pay our own expenses, we could not clearly see what this was for; but it occurred to us, after a great deal of deliberation, that it was a kind of bounty allowed by the city council to the agent for disposing of all vagrants who came within his reach, and that he had, through the force of habit, or in the confusion of his multifarious duties, mistaken us for persons of that description.

On our passage to Providence, the steam-boat touched at Newport, where one of our whalemen, who had made a raise of three dollars from the New York agent—in remembrance, he said, of a whaling voyage on which the old gentlemen had sent him a few years previously—privately notified us of his intention to "visit some of his friends up town." Before parting from him, he gave us his experience as a whaleman, and advised us not to be gulled by fair promises. He said he knew a thing or two about it; that he would sooner be in the penitentiary any time; and, if we had any regard for ourselves, we ought to turn our backs upon New Bedford, for it was the sink-hole of iniquity; that the fitters were all blood-suckers, the owners cheats, and the captains tyrants.

****

I have not the conscience to pass over in silence the disinterested generosity of the New Bedford fitter. His benevolence surpassed even that of the amiable old gentlemen in New York. When we first presented ourselves for inspection, he was a little bluff, to be sure, but that was only one of his good-natured peculiarities.

"Why," said he, surveying us with professional deliberation, "you are not the men I wrote for. I want stout, hard-fisted fellows, who ain't afraid to work. Such slim chaps as you won't do at all!"

Part IV—Excerpts from *Etchings of a Whaling Cruise*

"That's rather hard, sir; here we are without the means of getting back; and now, after the New York agent telling us you would take us, you say we won't do."

"What do I care about the New York agent?" replied the fitter. "It's his own look-out, and yours, if he don't send proper men. I'm not bound to take you at all; and I won't take you, if I don't like."

"Well, you'll pay our expenses back, then?"

At this the fitter laughed very heartily.

"No, no, my good fellows; can't do that. I see you don't understand this business. What do you weigh?"

We gave him our weight, but it did not seem to satisfy him exactly. He shook his head with a doubtful look, as much as to say he had no great respect for men who did not weigh considerably over our standard. He then punched us with his fist, shook us by the arms, and, after some further experiments by way of testing our muscular powers, told us what there was of us was pretty good, "but there *wasn't enough*." Directing us next to walk up and down his long store-room, he planted himself against a pile of boxes, and watched our gait with the practiced eye of a jockey about to make a speculation in horse-flesh. Apparently satisfied, he ventured the opinion that we might do; at all events, he would exert his influence in our behalf with the owners.

A clerk who sat in the counting-room, blowing his very soul through a cracked fife, was then directed to show us to old Captain R——'s boarding house. Here we found a most jovial company; not very select, but remarkably free and easy. Among others, I recollect Red Sandy, Blue John, Long-legged Bill, Big-foot Jack, Chaw-o'-tobacco Jim, Handsome Tom, and one of our steam-boat acquaintances, who had already obtained the *soubriquet* of Bully Clincher; besides four lively house-maids, whom the sailors called Mag, Moll, Bet, and Peg, and with whom they seemed to be on the most friendly terms.

Our fellow-boarders, when the fact became known that we were about to go to sea, entertained themselves with sundry jests at our expense, all of which we took with the utmost good humor. This completely disarmed them. We were shrewd enough to suspect their object, which, as we afterward learned, was to get us angry, and then, according to custom, give us a sound drubbing. Sailors have an inveterate dislike to young sprigs, who, when placed upon a level with them, assume airs of superiority. By guarding against this, we became great favorites. I must not omit, however, to men-

Part IV—Excerpts from Etchings of a Whaling Cruise

tion one of the initiatory movements. While standing at the door, the first evening after our arrival, we overheard the comments made upon ourselves and our mission.

"I say, Bill," said one, "there's a pair of bloody tars for you! They'll be slushin' down the t'gallant mast before long, or I'm out o' my reckoning."

"Ay, ay," replied Bill; "better they never was weaned, than go driftin' round the world in blubber hunter."

"Never mind," added another, "they'll wish themselves in the watch-house before two months."

With these and other remarks of the kind they amused themselves for some time, when one of the party, a regular old sea dog, with a tremendous quid of tobacco in his cheek, waddled up to us, and, staring us in the face, exclaimed,

"Well, cuss me if these ain't the lob-lolly boys wot sarved in one of my ships. I say, my lads, don't you know your old skipper? I'm Captain Bill Salt, wot used to larn you Lunars. Don't you know me?"

"No; you must be mistaken. We have never been to sea."

"Now I'm shivered if that ar'n't strange!" cried Captain Bill Salt; "if you ain't my lob-lolly boys, I never seed 'em."

"Nevertheless, we are not. B—— is my name, and W—— is my friend's."

"Well, just as good. You was both born to go to sea. Come, let's splice the main brace. Come along, shipmates! I'm agin' to give these 'ere young gentlemen the first lesson in Lunars."

Captain Bill Salt's manner was, to say the least of it, very friendly. We thought it best not to refuse his polite invitation. The sailors followed their comrade, who led the way to a chop-cellar a short distance from the boarding-house.

"Come, all hands, what'll you take? Don't be shy. What d'ye say, shipmates," addressing W—— and myself; "close-reef or sea-breeze?"

"Close-reef," said we, at a guess.

"Bravo!" cried Captain Bill, grasping each of us by the hand; "you'll see the stars yet! If you ain't sailors, it's the 'fects of eddecation or s'ciety, wot's all the same. Come, here's a toast:

> 'Be cheery, my lads! may your hearts never fail
> While the bold harpooneer is striking the whale!'

The toast was duly honored; and we discovered, when we emptied our glasses, that "close-reef" was something very strong. Big-foot Jack,

Part IV—Excerpts from *Etchings of a Whaling Cruise*

Chaw-o'-tobacco Jim, Handsome Tom, Red Sandy, and the rest of our jolly friends, then seated themselves and called for cigars. Captain Bill Salt told us to do likewise; and, taking out his pipe, he soon enveloped himself in a comfortable cloud of smoke. Without waiting for the ceremony of an invitation, he gave vent to the following ditty, a copy of which I afterward procured from him:

"PARTING MOMENTS.

"Farewell, my lovely Nancy,
  Ten thousand times adjeu!
I'm agoing for to cross the ocean
  In sarch of something new.
Come, change a ring wid me, my dear,
  Come, change a ring wid me;
And that will be my fond toaken
  When I am on the sea—
    When I am on the sea,
    And you don't know where I be.

Now one fond kiss, my Nancy dear,
  Now one fond kiss for me,
Before I go for to begin
  To roam upon the sea.
And hear this secret of my heart:
  Wid the best of my good-will,
Be where it may, this poor body,
  Is yourn, sweet Nancy, still—
    Is yourn, sweet Nancy, still,
    Wid the best of my good-will."

  This song elicited the most rapturous applause. Captain Bill then spun us some tough yarns, while the company slipped out one by one. As we were about to leave, the bar-keeper called us aside, and politely requested my friend and myself to pay the reckoning, assuring us that it was customary, when young gentlemen were about to go on a voyage, to treat all hands. We accordingly gave him our last cent, and were not a little edified at the cool manner in which Captain Bill Salt witnessed the operation. Though our confidence in that eccentric individual was a little shaken, we took the whole proceeding as a very good joke, and laughed to think how cleverly we had been gulled. Thus ended our "first lesson in Lunars."
  Our friend, the fitter, was a most accommodating man. With a delicate appreciation of our pecuniary embarrassments, he paid our board, fur-

Part IV—Excerpts from Etchings of a Whaling Cruise

nished us with every little luxury we wanted, lent us his pleasure-boat to sail in, told us he would make our expenses all right with the owners, and gave us a great deal of fatherly advice about our conduct at sea. In addition to all this kindness, he considerately provided us with chests and sea-clothes at a terrible sacrifice, being at least ten per cent cheaper than we could get them elsewhere. Besides, the mere fact of his crediting total strangers seemed so generous, so confiding, so high-minded!

The only vessel about to sail immediately was the barque *Styx*, of Fair Haven. Through the exertions of our excellent friend, the fitter, the owners, apparently with great reluctance, agreed to take us. They told us the vessel was well fitted; better, in fact, than any vessel we could find. One of them, an old Quaker, assured us no whaler had ever sailed from New Bedford or Fair Haven as well fitted; he had attended to it all himself, and, we might depend upon it, we would live in style. The captain, we learned from them, was a young man, pretty strict in his discipline, but a fine, generous fellow. He would treat us well, and give us plenty to eat; and, if we made ourselves useful, he would be very kind to us. He was a first-rate whaleman, and no doubt we would make a good voyage, and come home in a year or a year and a half with lots of money due to us. The vessel was a hundred and forty-seven tons burden, and calculated to hold a thousand barrels of oil. We were to receive the ordinary lay of green hands, being, as we were told, the one hundred and thirtieth part of the oil taken. There was provision enough on board to last for twenty-seven months, so that, if not. successful, there was no danger of our starving. We were to have what clothes we needed out of the slop-chest at the New Bedford prices. The shipping articles were then presented to us, and we signed them without exhibiting any such ungentlemanly want of confidence in the representations of the owners as to read the contents; besides, we were afraid, as they had accepted us so reluctantly, some difficulty might arise by which we would be deprived of the pleasure of performing a voyage under such pleasant auspices. The signing of the articles we regarded as a sort of security.

With sanguine hopes and enthusiastic dreams of adventure we bade good-by to our New Bedford friends and embarked. The *Styx* lay in the middle of the Acoshnet River, opposite the town of New Bedford.

At 2 p.m. all hands were called to the windlass, and we weighed anchor. A light breeze slowly wafted us out into Buzzard's Bay. The shipping at the New Bedford wharf became gradually indistinct, and the houses looked misty in the distance. It was a beautiful Sabbath afternoon. The

Part IV—Excerpts from *Etchings of a Whaling Cruise*

church bells were tolling a melancholy farewell; and I shall never forget the look W—— gave me as he pointed to the receding shores, and observed, in a melancholy tone, "I have unhappy thoughts. It seems to me those familiar sounds call us back. But we are too late; it is useless to repent now." My feelings were touched; the whole past was before me in a moment: friends, brothers, sisters, all! I would have given all I ever hoped to possess to retrace a few hours of my life.

****

Toward evening the captain came on board in a pilot-boat, and took charge of the vessel. I had not seen him before, and of course felt a curiosity to know what sort of a looking man he was. The owners had spoken in such glowing terms of him that, I must confess, he did not altogether realize my expectations. His personal appearance was any thing but prepossessing. Picture to yourself a man apparently about thirty-five years of age, with a hooked nose, dark crop hair, large black whiskers, round shoulders, cold blue eyes, and a shrewd, repulsive expression of countenance; of a lean and muscular figure, rather taller than the ordinary standard, with ill-made, wiry limbs, and you have a pretty correct idea of Captain A——. He wore a broad brimmed Panama hat, turned up at the sides, a green roundabout, a pair of dirty duck pantaloons, very wide at the bottom, and slip-shod shoes, which had evidently done service for two or three voyages. He walked the quarter-deck with his hands in his pockets, his eyes down, and his lips firmly compressed. Altogether he had a sneaking, hang-dog look that was not very encouraging to those destined to be subject to his will during a year's cruise, or perhaps longer. When he gave orders, it was in a sharp, harsh voice, with a vulgar, nasal twang, and in such a manner as plainly betokened that he considered us all slaves of the lowest cast, unworthy of the least respect, and himself our august master.

Night closed upon us with rough and cloudy weather. By morning we had a heavy, chopping sea, and began to experience all the horrors of sea-sickness. The mate, a stout, bluff-looking Englishman, with a bull neck, kept us in continual motion, and gave us plenty of hard work to do, clearing up the decks, bracing the yards, stowing down the loose rubbish, and otherwise making the vessel tidy and ship-shape. He bellowed forth his orders to the men in the rigging like a roaring lion, yelled and swore at the "green hands" in the most alarming manner, and pulled at the ropes as if determined to tear the whole vessel to pieces. The loungers or "sogers" had no

Part IV—Excerpts from Etchings of a Whaling Cruise

chance at all with him; he actually made them jump as if suddenly galvanized. For the seasick he had no sympathy whatever.

"Stir yourselves; jump about; pull, haul, work like vengeance!" he would say, in the bluff, hearty voice of a man who appeared to think sickness all folly; "that's the way to cure it. You'll never get well if you give up to it. Tumble about there! Work it off, *as I do!*"

To the haggard, woe-begone landsmen, who staggered about groaning under their afflictions, this sounded very much like mockery. For my part, I thought the mate a great monster to talk about sickness, with a face as red as a turkey-cock's snout.

After a day of horrors such as I had never spent before, we were permitted to go below for the night. Our condition was not improved by the change. The forecastle was black and slimy with filth, very small, and as hot as an oven. It was filled with a compound of foul air, smoke, sea-chests, soap-kegs, greasy pans, tainted meat, Portuguese ruffians, and sea-sick Americans. The Portuguese were smoking, laughing, chattering, and cursing the green hands who were sick. With groans on one side, and yells, oaths, laughter and smoke on the other, it altogether did not impress W—— and myself as a very pleasant home for the next year or two. We were, indeed, sick and sorry enough, and heartily wished ourselves ashore.

****

Sea-sick and harassed after a hard day's work, we had gladly availed ourselves of a few hours' respite from duties so laborious. The mate came to the scuttle, and, with half a dozen tremendous raps, roared at us to bear a hand. "Tumble up, every mother's son of you, and take in sail. Out with you, green hands and all. We won't have any sick aboard here. You didn't come to sea to lay up. No groaning there, or I'll be down after you. D'ye hear the news down below? Tumble up! tumble up, my lively hearties!"

There was no refusing so peremptory a command as this, little as we liked it. Without exactly *tumbling up*, we contrived, with some difficulty, to gain the deck, for the vessel pitched so violently that few of the green hands could keep their feet under them. I shall never forget the bewilderment with which I looked around me. We were in the Gulf Stream, enshrouded in darkness and spray. The sea broke over our bows, and swept the decks with a tremendous roar. Momentary flashes of lightning added to the sublimity of the scene. When I looked over the bulwarks, it seemed to me that the horizon was flying up in the clouds and whirling round the vessel by turns, and the clouds, as if astonished at such wild pranks, appeared to be shaking their dark heads backward and forward over the horizon. I looked aloft, and there the sky was sweeping to and fro in a most unaccountable manner. The vessel went staggering along, creaking, groaning, and thumping its way through the heavy seas.

    I grasped the first rope I could get hold of, and held on with the tenacity of a drowning man. For a few moments I could do nothing but gasp for breath, and wipe the salt water out of my eyes with one hand while I held on with the other. The confusion of voices and objects around me, the tremendous seas sweeping over the decks, and the flapping of the sails, impressed me with the belief that we were all about to be lost. I kept my grasp on the rope, thinking it must be fast to something, and, if the ship foundered, I should at least be sure of a piece of the wreck. As for my comrade W——, I supposed he was still on board, and called for him with all my might; but the wind drove my voice back in my throat. While standing in this unpleasant predicament, the mate came rushing by, shouting to the green hands to "tumble up aloft, and lay out on the yards!" Aloft such a night, and for the first time! Was the man mad? The very idea seemed preposterous. Presently he came dashing back, thundering forth his orders with the ferocity of a Bengal tiger. "Up with you! Every man tumble up! Don't stand gaping like a parcel of boobies! Aloft there, before the sails are blown to Halifax!" Knowing how useless it would be to remonstrate, and believing I might as well die one way as another, I sprang up on the weather bulwark and commenced the terrible ascent. The darkness was so dense that I could scarcely see the ratlins, and it was only by groping my way in the wake of those before me, that I could at all make out where I was going. A few accidental kicks in the face from an awkward fellow who was above me, and a punch or two from another below me, convinced me that I was in company, at all events. How I contrived to drag myself over the foretop, I do not well remember. By a desperate exertion, however, I succeeded, and holding on to every rope I could get hold of with extraordinary tenacity, I at length found myself on the foot-rope, leaning over the yard and clinging to one of the reef-points, fully determined not to part company with that in spite of the captain, mate, or whole ship's company. "Haul out to leeward!" roared somebody to my right; "knot away!" This was all Greek to me. A sailor close by good-naturedly showed me what I was to do, and having knotted my reef-point, I looked down to see what was the prospect of getting on deck again. The barque was keeled over at an angle of forty-five

Part IV—Excerpts from Etchings of a Whaling Cruise

degrees, plunging madly through the foam, and I could form no idea of the bearings of the deck. All I could see was a long dark object below, half hidden in the raging brine. My right-hand neighbor gave me a hint to get in out of the way, which required no repetition, for I found my situation anything but pleasant. By the time I reached the fore-top my head was pretty well battered, and my hands were woefully skinned and bruised, the sailors having made free use of me to accelerate their downward progress.

I found, on gaining the forecastle, that my friend W—— had passed through the ordeal in safety. We said nothing, but looked our unqualified disapprobation of such a life. The Portuguese, to make matters still worse, laughed heartily at the sorry figure we cut, and told us all this "was nothing to what we'd see yet."

Next day the green hands, including my friend and myself, looked haggard enough. We were all dreadfully sea-sick. Our fare was by no means inviting under such circumstances. For breakfast we had an abominable compound of water, some molasses, and something dignified by the name of coffee, with hard biscuit and watery potatoes; for dinner pork, salt beef, and potatoes; and for supper, a repetition of the biscuit and potatoes, with boiled weeds and molasses as a substitute for tea and sugar. It was perfectly amazing the voracity with which the Portuguese devoured this fare. Had they whetted their appetites for months on raw corn they could not have swallowed such food as was now before them with more relish. I must confess, their digestive powers excited my envy as well as my astonishment. It made me despair to see them eat. I would have given all I expected to make during the voyage to possess their swinish relish for food. However, before the expiration of two months, I had reason to change my tune. I would have given twice as much to get rid of my appetite!

****

The captain deliberately stalked the quarter-deck, exulting in the "pomp and circumstance" of his high and responsible position. Every step he took bespoke the internal workings of a man swelling with authority. The proud glance of his eye; the severe frown of his heavy eyebrows; the haughty curl of his lip; even the peculiar twist of his long, nasal protuberance seemed to say, "Behold, and wonder! I stand before you arrayed in a halo of glory. I am commander of the great barque *Styx*! Authority is mine! Look upon me, all ye who have eyes to see, and tremble, all ye who have ears to hear!" With his hands stuck in his breeches pockets, he then

Part IV—Excerpts from *Etchings of a Whaling Cruise*

approached the break of the quarter-deck, and, straddling out his legs to guard against lee-lurches, asked if all hands were present. One of the officers replied in the affirmative.

The scene was at once grotesque and impressive. Fourteen men, comprising the whole crew, were huddled together in the waist, at the starboard gangway. Of these four were Portuguese, two Irish, and eight Americans; and certainly a more uncouth-looking set, including my friend and myself, never met in one group. The Portuguese wore sennet hats with sugar-loaf crowns, striped bed-ticking pantaloons patched with duck, blue shirts, and knives and belts. They were all barefooted, and their hands and faces smeared with tar. On their chins they wore black, matted beards, which had apparently never been combed. The color of their skin was a dark, greenish-brown, if the reader can imagine such a color, and was calculated to create the impression that they never made use of soap and water. The variety of dress in which the rest of the crew were habited was fully as striking as that of the Portuguese. Some wore Scotch caps, duck trowsers, red shirts, and big horse-leather boots; others, tarpaulin hats, Guernsey frocks, tight-fitting cloth pantaloons, and red neckerchiefs. Several were bareheaded and barefooted, having lost their hats and shoes in the late gale. All the green hands, which included most of the Americans and the two Irishmen, were still cadaverous and ghastly after their sea-sickness, and not more than two had yet entirely "squared accounts with old Nep." Altogether we were the most extraordinary looking set of half-sailor nondescripts possible to conceive. Thus situated, and thus equipped for sea life, we stood gaping at the captain in silent admiration.

The captain, after considerable deliberation, and a great show of contempt toward every body within range of his visual rays, then addressed us in a sharp nasal voice, fixing his eyes upon each man alternately. I had listened to many speeches, but never to one more pointed than this. No doubt he will be surprised to find it literally reported.

"I suppose you all know what you came a whaling for if you don't, I'll tell you. You came to make a voyage, and I intend you shall make one. You didn't come to play; no, you came for oil; you came to work." [Here he took a turn on the quarter-deck while concentrating his ideas for another burst of eloquence, amused himself in an undertone, partly addressed to himself individually, and partly to the mate, by letting us know that it should be "a greasy voyage, and a monstrous greasy one too."]

"You must do as the officers tell you, and work when there's work to

Part IV—Excerpts from Etchings of a Whaling Cruise

be done. We didn't ship you to be idle here. No, no, that ain't what we shipped you for, by a grand sight. If you think it is, you'll find yourselves mistaken. You will that—some, I guess." [Here he lost the idea, or, to use a more expressive phrase, "got stumped."] "I allow no fighting aboard this ship. Come aft to me when you have any quarrels, and *I'll* settle 'em. *I'll* do the quarreling for you—*I* will." [Another turn on the quarter-deck.] "If there's any fighting to be done, I want to have a hand in it. Any of you that I catch at it, 'll have to FIGHT ME." [A frightful doubling up of the fists, and a most ferocious gnashing of the teeth.] "I'll have no swearing, neither. I don't want to hear nobody swear. It's a bad practice—an infernal bad one. It breeds ill will, and don't do no kind o' good. If I catch any one at it, damme, I'll flog him, that's all." [A nod of the head, as much as to say he meant to be as good as his word.] "When it's your watch below, you can stay below or for'ed, just as you please. When it's your watch on deck, you must stay on deck, and work, if there's work to be done. I won't have no skulking. If I see sogers here, I'll soger 'em with a rope's end. Any of you that I catch below, except in cases of sickness, or when it's your watch below, shall stay on deck and work till I think proper to stop you." [A stride or two aft, and a glance to windward.] "You shall have good grub to eat, and plenty of it. I'll give you vittles if you work; if you don't work, you may starve. Don't grumble about your grub neither. You'd better not, I reckon." [A mysterious shake of the head, which implied a vast deal of terrific meaning.] "If you don't get enough, come aft and apply to me. *I'm* the man to apply to; *I'm* the captain." [Here he surveyed himself with a look of exultation, which seemed to say that he was not only the captain, the *very* man to whom he had special reference, but that it was a source of infinite satisfaction to him to *be* the captain.] Now, the sooner you get a cargo of oil, the sooner you'll get home. You'll find it to your interest to pay attention to what I say. Do your duty, and act well your part toward me, and I'll treat you well; but if you show any obstinacy or cut up any extras, I'll be d—d if it won't be worse for you! Look out! I ain't a man that's going to be trifled with. The officers will all treat you well, and I intend you shall do as they order you. If you don't, *I'll* see about it." [Three or four strides to and fro on the quarter-deck, and a portentous silence of five minutes.] "That's all. Go for'ed, where you belong!"

Had the captain made good all his promises, we would have had no just cause for complaint; but we soon discovered that his speech was merely designed to intimidate us. From that time forth we had the poorest fare,

Part IV—Excerpts from *Etchings of a Whaling Cruise*

and in the scantiest quantities. The owners had given us positive assurance that there never had sailed from that port a vessel better fitted in every respect. For their misrepresentations, we heartily wished them a berth in their own barque, believing that the severest punishment that could be inflicted upon them. A month's trial at it would make them exercise more humanity toward their fellow-creatures.

Next in the routine of business was the choosing of watches. We were all called to the waist that evening, and examined like a parcel of bullocks about to be butchered. The mate and second mate made the selections. Among others, I was chosen for the larboard or mate's watch, and my friend for the star-board or second mate's watch.

The watch on deck was then set to work on the whaling gear. Our duties from that time till we arrived on the western whaling ground were, working ship, grinding harpoons, spades, lances, boarding knives, &c, making deck brooms, washing decks every morning, clearing the rubbish away every afternoon, stowing away loose casks, steering and standing mastheads. Whenever the weather was fine we lowered the boats and practiced at pulling, backing, and all the manceuvers necessary in the capture of a whale. All this severe labor was very hard upon those who had not been accustomed to great physical exertion.

****

July 27th.—I had afternoon watch below, and had turned in to forget my troubles in sleep. About two o'clock I was roused by the steward, who informed me that W—— had suddenly fallen upon the deck in a fit of convulsions. I immediately sprang up the ladder and ran aft. Language can not depict the shocking spectacle that met my eyes. There was my bosom friend, sitting up against one of the scuttle-butts, his shirt open, his hat lying on the deck, and his eyes ready to start from their sockets. The captain stood by him, holding him by the hand. I felt sick and giddy, when W—— stared at me with the vacant gaze of an idiot. Bursting into a wild laugh, he attempted to spring up. It was a fearful laugh—a laugh that rang like a death-knell in my ears. I grasped him by the hand; the terrible thought struck me that he had gone mad! His voice was wild and unnatural, and his whole appearance awful in the extreme. Gazing vacantly in my face, he burst into tears, and sobbed as if his heart would break. I called him by name; I implored him to speak to me. Without noticing my appeals, he turned to the captain and inquired my name. Upon receiving an answer,

Part IV—Excerpts from Etchings of a Whaling Cruise

he begged me, in the most piteous tones, to convey a message home to his mother, that he never should see her again.

"Before another hour," he said, "I shall be food for the sharks. O God, must I die so soon? Am I never to see home again? I have kind, good parents; tell them I died thinking of them. It is horrible—horrible to be thrown overboard in a sack."

No effort to console him had the slightest effect. The fearful idea that he was about to be devoured by the sharks seemed to drive him mad. He raved of strange things which he had seen at the masthead; talked incoherently of birds with beautiful plumage, curiously-formed fishes, and called upon us wildly to save him from the sharks. It was a scene of horror that I shall never forget.

When he became somewhat composed, one of the hands, assisted by myself, carried him forward to the forecastle, and laid him in his berth. For three hours he lay in a trance, with his eyes wide open, not moving a muscle. He looked like one that was dead.

It appeared, from the statements of the watch on deck, that he had just come down from the masthead, where the rays of the sun poured down with an intense heat. On reaching the deck, he walked aft toward the captain, who was parading the quarter-deck. After passing the break of the deck he stood still, and while in the act of addressing the captain, fell down in convulsions. From all these circumstances, and from the fact that he was not subject to fits, it was quite evident that it was a sunstroke. He had suffered severely from sea-sickness, and was greatly debilitated. A burning sun beating down upon his head for two hours could very easily have produced the terrible effects described.

I thought it very hard that a man, really suffering from illness, should be compelled by the captain to stand two hours a day at the mast-head. It was, in this case at least, little better than murder. W—— never recovered from the effects of this fearful affliction. Better, far better would it have been for him, had he fallen from his post and found a watery grave. There are things connected with this event that weigh heavily upon my heart; things not rudely to be touched affections tried and hearts broken.

It is needless to dwell upon his sufferings during the remainder of his stay on board the ship. The Portuguese were mere brutes, and, with two or three exceptions, the rest of the crew were little better. Sympathy for the sick was a weakness unknown to them. No temptation would induce them to refrain from smoking, swearing, and blackguarding. I attempted to pur-

Part IV—Excerpts from *Etchings of a Whaling Cruise*

chase peace by giving them my clothes, but my exertions were of no avail. I saw that it was useless to expostulate, and finding that the noise increased W——'s malady, I appealed to the captain to exert his influence over them. His reply was characteristic, and just such as I might have expected had I known him better. "He had nothing to do with the forecastle. The Portuguese, as well as the Americans, were at liberty to do as they pleased in it. He had no control over them after they went below. W—— had no business coming to sea to get sick, and be a trouble to all on board. He had seen such fellows before, and would not put himself out of his way to pamper to their wants. Now that he was in a scrape, let him make the best of it, and not trouble folks with his complaints. If he wanted medicine, he might have it, and that was all that could be done for him."

Where such an example was set by the captain, I could not expect the crew to do otherwise than follow it. For FIFTY-TWO days W—— lay in the forecastle, suffering such tortures of body and mind as can not be described. The captain gave him to understand that he should not leave the vessel the whole voyage; he might die in the forecastle, for what he cared. During all this time, my unfortunate comrade had nothing to eat but hard biscuit, and occasionally a piece of butter about the size of a dollar; so reduced was he that nothing else allowed the crew would remain on his stomach. The hot, close atmosphere of the forecastle, rendered still more suffocating by the fumes of old pipes and bad cigars, was not very well calculated to promote his recovery.

It would be difficult to give any idea of our forecastle. In wet weather, when most of the hands were below, cursing, smoking, singing, and spinning yarns, it was a perfect Bedlam. Think of three or four Portuguese, a couple of Irishmen, and five or six rough Americans, in a hole about sixteen feet wide, and as many, perhaps, from the bulk-heads to the fore-peak; so low that a full-grown person could not stand upright in it, and so wedged up with rubbish as to leave scarcely room for a foothold. It contained twelve small berths, and with fourteen chests in the little area around the ladder, seldom admitted of being cleaned. In warm weather it was insufferably close. It would seem like exaggeration to say, that I have seen in Kentucky pig-sties not half so filthy, and in every respect preferable to this miserable hole: such, however, is the fact.

In this loathsome den, the Portuguese were in their element, reveling in filth, beating harsh discord on an old viola, jabbering in their native language, smoking, cursing, and blackguarding. Their chief recreation, how-

Part IV—Excerpts from Etchings of a Whaling Cruise

ever, was quarreling, at which they were incessantly engaged. Nor was it confined to week-days, for not the slightest regard was paid to the Sabbath. The most horrible profanity was indulged in, and to an excess that was truly revolting. They did not seem aware even of the existence of a Supreme Being. And yet these Christians chattered a paternoster over their beads every night! What mockery!

****

As soon as we arrived on the western whaling ground, boat watches were set. In a small vessel like the *Styx*, with three boats, besides a spare boat aft, there are usually three watches, consisting of the larboard, starboard, and waist boat's crew. Each watch is under the command of a boat-steerer after sail is shortened, which is generally about sundown. In our watches there were four men, and the boat-steerer. The mate and second mate sleep all night, and remain on duty all day. The alternate hours of duty and rest with the crew are arranged thus. Say the larboard and starboard boat's crews go below after sail is taken in; the waist boat's crew remains on deck till ten o'clock, when it is relieved by the larboard boat's crew, and turns in till the hands are called in the morning. The watch then on deck is relieved at one by the starboard boat's crew, which remains on deck till all below are called in the morning. The starboard watch then has forenoon watch below, the larboard the afternoon, and the waist boat's crew all day on deck.

In making a passage, there are but two watches, the larboard and starboard, which are headed by the first and second mate, who take the same hours of rest allowed the crew.

August 3d.—We had now prepared all the whaling gear, and were daily on the look-out for whales.

August 5th.—The boats were lowered for black-fish. I took my place, for the first time, at the aft oar in the waist boat. After rowing about two miles, we came up with the school. It was an unusually large one, but the day was so calm that they were very shy. We made several unsuccessful attempts to get a dart at them, and continued the chase for six or eight hours under a burning sun. I was pretty well tired of my oar by the time we turned toward the vessel. The Portuguese consoled me with the remark, that I had not begun to see "a hard pull yet" and enjoyed my cadaverous looks with great satisfaction.

From seven till nine o'clock we usually spent on deck, amusing our-

Part IV—Excerpts from *Etchings of a Whaling Cruise*

selves at the various pastimes common among sailors. When the weather permitted, we had dancing, singing, and spinning yarns. The Portuguese had a guitar, or viola, as they called it, with wire strings, upon which they produced two or three melancholy minors, accompanying their performance with a harsh, unmusical chant. Four of them formed couples, and while one of the by-standers played the guitar, those forming the set moved backward and forward like hyenas in a cage, pawing the deck with their feet, and using their fingers by way of castanets; all chanting, in a whining tone, two or three monotonous notes, which they repeated till it became fairly distracting. While the Portuguese amused themselves in this way, the American portion of the crew had songs, yarns, and dances after their own fashion. As all human enjoyments are comparative, so many an hour of real pleasure was thus passed on board the *Styx* by myself and others, who had seen worse times since we had left New Bedford.

****

To some readers, who derive their ideas of things aboard ship from sea novels, in which the valor of the heroes consists in a heroic contempt of all authority, and a superabundance of impertinence, it may seem that to submit tamely to the overbearing bullying of a brute, without retort or resentment, shows a want of manly spirit. I would ask, what is to be done in such cases? A man has no right to strike his commander, however well justified he may be in so doing, according to our notions of right and wrong. Nor must he use language that can be termed insolent or mutinous. This might do ashore, where one man can meet another upon equal terms; but it can not be carried out at sea.

****

August 16th.—Chased a *school* of whales all day. At 6 o'clock p.m. their spouts were seen about two miles off the lee bow. The larboard and starboard boats, headed by the captain and the mate, were lowered. At 10 p.m. the boats came alongside with a twenty-barrel whale in tow. All hands set to work rigging up the cutting tackle, and getting the try-works ready.

The appearance of this, our first whale, was hailed by a general cheer. After the watches were set, and the decks cleared, I had an opportunity of examining our prize. It was about thirty-five feet in length, of a rather light color, and had a strong, disagreeable smell of oil. Though considered a very small whale, its proportions seemed gigantic enough to me. It was surrounded by sharks eagerly awaiting their prey.

****

I was much amused at the remarks of the "downeaster," suggested by the novel appearance of our first whale. I observed him, as he leaned over the monkey-rail, gazing steadfastly at the whale, while he muttered something to himself which I could not hear.

"Well, Mack," said I, "what's your opinion of whales?"

"Why, I was jest a thinkin' it's a considerable sort of a fish. They ain't got fish like that up the Kennebeck."

"I guess not. Still it is nothing like so large as the whale Jonah swallowed."

"By gosh!" shouted Mack, laughing, "if his'n was bigger than that, I'll be durned if the flukes didn't tickle his throat, if he was as sea-sick as I was a spell ago."

"Do you think whales are fish?" said I, rather balked in my attempt to quiz him.

"Why, some folks says whales isn't fish at all. I rayther calculate they are, myself. Whales has fins, so has fish; whales has slick skins, so has fish; whales has tails, so has fish; whales ain't got scales on 'em, neither has catfish, nor eels, nor tadpoles, nor frogs, nor horse-leeches. I conclude, then, whales *is* fish. Every body had oughter call 'em so. Nine out of ten *doos* call 'em fish. If whales live on small fish, they'd drive a smashin' business up the Kennebeck. I never see none up thar'. If I was a whale, I'd try them diggins. There ain't better fodder for whales no whar'. This may be a good place, for all I know; but it looks dreadful blue and lonesome. I'd want to be in fresh water, if I was a whale; and then, if I wanted to season the vittles Natur' gave me, I'd pile the salt on rayther more moderate. I'd salt 'em to suit me. I don't like to be forced to eat salt vittles now, and I ain't a whale. Whales is cannibals. I've a bad opinion of 'em myself. I don't like the looks of 'em, no how. Gosh! What a jaw! I'd rayther let 'em be, and do business on a smaller scale. Folks that doos business on a small scale ain't so likely to git bu'st. Fishin's a fishin'. I like fishin' as well as any body; but catchin' of whales is a leetle too extensive. It's or fully alarmin' work. I don't want to be swallered jest yet; not in the whalin' line, I don't!"

At daylight next morning all hands were called, and set to work upon the whale. A brief description of the process of procuring the oil may not

Part IV—Excerpts from *Etchings of a Whaling Cruise*

be uninteresting. The blubber varies from four to ten inches in thickness. It is cut from the whale in layers about three feet wide, which run from the head to the flukes, in a spiral form. After the blubber and flukes are hoisted on board with a large tackle attached to a pendant in the main-top, the boat-steerers cut them in sizes sufficiently small to fit snugly in the blubber-room, an apartment in the main hold. The try-works are then cleaned out, and got in readiness for boiling. Two or three hands are stationed in the blubber-room with short spades, whose duty it is to cut up the large pieces of blubber called blanket pieces into blocks or pieces about a foot and a half long and six inches wide. The blubber is then minced into thin slices, and cast into the boilers; a fire started, and the first batch of oil obtained : the crisped pieces of blubber are used for fuel. The hot oil is strained into a large copper cooler, where it is permitted to settle till the boilers are again ready to be emptied. It is then strained into casks, and kept on deck till quite cool, when it is stowed down in the casks in the hold by means of a hose.

A "trying out" scene is the most stirring part of the whaling business, and certainly the most disagreeable. The try-works are usually situated between the foremast and the main hatch. In small vessels they contain two or three large pots, imbedded in brick. A few barrels of oil from the whale's case, or head, are bailed into the pots before commencing upon the blubber. Two men are standing by the mincing horse, one slicing up the blubber, and the other passing horse pieces from a tub, into which they are thrown by a third hand, who receives them from the hold. One of the boat-steerers stands in front of the lee pot, pitching the minced blubber into the pots with a fork. Another is stirring up the oil, and throwing the scraps into a wooden strainer. We will now imagine the works in full operation at night. Dense clouds of lurid smoke are curling up to the tops, shrouding the rigging from the view. The oil is hissing in the try-pots. Half a dozen of the crew are sitting on the windlass, their rough, weather-beaten faces shining in the red glare of the fires, all clothed in greasy duck, and forming about as savage a looking group as ever was sketched by the pencil of Salvator Rosa. The cooper and one of the mates are raking up the fires with long bars of wood or iron. The decks, bulwarks, railing, try-works, and windlass are covered with oil and slime of black-skin, glistering with the red glare from the try-works. Slowly and doggedly the vessel is pitching her way through the rough seas, looking as if enveloped in flames.

"More horse pieces!" cries the mincer's attendant.

Part IV—Excerpts from Etchings of a Whaling Cruise

"Horse pieces!" echoes the man in the waist.

"Scraps!" growls a boat-steerer.

By-and-by the captain comes up from the cabin to see how things are progressing. He peeps into the pots, and observes, in a discontented tone, "Why don't you keep that 'ere oil stirred. It's all getting black." Then he takes a look into the mincer's tub: "That won't do! Make Bible leaves of 'em." Then he looks at the men on the windlass: "Hey! all idle? Give these fellows something to do. We can't have idlers about now."

Having delivered himself of these sentiments, he goes back to his snug nest in the cabin. The idlers resume their places, and entertain themselves spinning yarns, singing songs, &c, and calculating the time by the moon. About the middle of the watch they get up the bread kid, and, after dipping a few biscuit in salt water, heave them into a strainer, and boil them in the oil. It is difficult to form any idea of the luxury of this delicious mode of cooking on a long night-watch. Sometimes, when on friendly terms with the steward, they make fritters of the brains of the whale mixed with flour, and cook them in the oil. These are considered a most sumptuous delicacy. Certain portions of the whale's flesh are also eaten with relish, though, to my thinking, not a very great luxury, being coarse and strong. Mixed with potatoes, however, like "porpoise balls," they answer very well for variety. A good appetite makes almost any kind of food palatable. I have eaten whale-flesh at sea with as much relish as I ever ate roast-beef ashore. A trying-out scene has something peculiarly wild and savage in it; a kind of indescribable uncouthness, which renders it difficult to describe with any thing like accuracy. There is a murderous appearance about the blood-stained decks, and the huge masses of flesh and blubber lying here and there, and a ferocity in the looks of the men, heightened by the red, fierce glare of the fires, which inspire in the mind of the novice feelings of mingled disgust and awe. But one soon becomes accustomed to such scenes, and regards them with the indifference of a veteran in the field of battle. I know of nothing to which this part of the whaling business can be more appropriately compared than to Dante's pictures of the infernal regions. It requires but little stretch of the imagination to suppose the smoke, the hissing boilers, the savage-looking crew, and the waves of flame that burst now and then from the flues of the furnace, part of the paraphernalia of a scene in the lower regions.

Of the unpleasant effects of the smoke I scarcely know how any idea can be formed, unless the curious inquirer choose to hold his nose over the smoking wick of a sperm oil lamp, and fancy the disagreeable experiment

Part IV—Excerpts from *Etchings of a Whaling Cruise*

magnified a hundred thousand fold. Such is the romance of life in the whale fishery. I have thus endeavored to describe a trying-out scene; and I hope, with the aid of a drawing taken on the spot, my hasty sketch will not be altogether unintelligible.

We saw, during our cruise on the western ground, great numbers of black-fish, grampus, porpoises, and jumpers; and caught in abundance dolphins, albacore, bonitos, and skip-jacks, which are all dry, and not to be compared with bay-fish.

****

In a journal of this kind, containing miscellaneous gatherings of every description, I can not well omit a sketch of what, in nautical phraseology, is termed "a gam." When two whalers meet on any of the whaling grounds, it is usual to have "a gam," or mutual visit, for the purpose of interchanging the latest news, comparing reckoning, discussing the prospect of whales, and enjoying a general chit-chat.

While our barque lay off Terceira, we one evening spoke a brother whaler. About four o'clock, when the decks were cleared up, the waist-boat was lowered, and we went on board with the captain. A crew from the stranger returned to the barque with our boat. After supper we had a social smoke. The musician of the ship was then called upon for a song. Seating himself comfortably on the fore-hatches, he cleared his throat, and gave us to understand, by way of a prelude, that he was a very indifferent singer. "He used to know some bang-up songs, but, some how, he had forgotten them all." This, of course, only served to whet our curiosity, and draw forth renewed calls for a song. Tom was a first-rate singer. Everybody knew Tom could sing. It was no use to deny it; Tom must sing! Pressed on all sides, Tom stuck his pipe in the galley, and scratched his head to rub up the musical organs. He then assured us that he once knew a great many songs.

"Come, Tom!" cried a chorus of voices, "give us 'Captain Bunker.'"

"Well, if I must, I must; here goes for 'Captain Bunker.'"

Tom then gave us the following whaling ditty. As it is a good specimen of sea-spun poetry, I give it without alteration:

"CAPTAIN BUNKER.

"Our captain stood upon the deck,
　A spyglass in his hand,
A viewing of those gallant whales
　That blowed at every strand.

Part IV—Excerpts from Etchings of a Whaling Cruise

> Get your tubs in your boats, my boys,
>   And by your braces stand,
> And we'll have one of those gallant whales,
> Hand, boys, over hand!
> *Chorus.* So be cheery, my lads! let your hearts never fail
>   While the bold harpooneer is a striking of the whale!
>
> "'Overhaul, overhaul!
>   Your davit-tackles fall,
> Till you *land* your boats in the sea
>   One and all!
> Our waist-boat got down,
>   And *of course* she got the start:
> 'Lay me on, Captain Bunker,
>   I'm h—l for a long dart!'
>   So be cheery, &c.
>
> 'Our first mate he struck,
>   And the whale he went down;
> The captain he stood by
>   All ready for to bend on;
> Which caused the whale to vomic,
>   And the blood for to spout:
> In less than ten minutes
>   He rolled both fins out!
>   So be cheery, &c."

Great applause was awarded Tom for the creditable manner in which he had acquitted himself. But singing was not altogether Tom's forte. According to the representations of his shipmates, he was "death on the fiddle!" The unanimous requests of the party were not to be resisted. The fiddler was compelled to play; and, while the two high functionaries aft were discussing matters of grave and momentous import, we entertained ourselves dancing "shindys" to the lively notes of Tom's fiddle. Those who could strike their heels together in the best time, go the double-shuffle with the greatest activity, and tire down their comrades, were of course the best dancers.

****

During our stay at the Island of Terceira, Smith, who had been off duty nearly the whole of our cruise up to that date, asked permission to go ashore. It was well understood by the captain and officers that he intended to desert, for he made no secret of it; and even went so far as to tie up a

Part IV—Excerpts from *Etchings of a Whaling Cruise*

bundle of clothes in his handkerchief, and carry it aft under his arm. Thinking this a good opportunity to get rid of him, without the expense of putting him ashore at the Villa Orta, which would not be less than forty dollars, the captain willingly gave him permission, telling him, as a matter of form, to be down at the boatlanding by sundown. Smith bid us all good-by, and was taken ashore in the waist-boat. The last I saw of him at that time, he was slowly dragging his emaciated limbs up the rocks.

On our arrival at the Villa Orta a week after, I was surprised to find Smith down at the landing, shaking hands with his old shipmates. It appeared that the vice-consul at Angra, to whom he had appealed, finding him in a destitute condition, had sent him over to Fayal in a fishing-smack, where he arrived a few days in advance of the *Styx*. There he made his complaint to the consul, who, of course, as is customary with consuls who have dealings with the masters of vessels, would have nothing to do with him. Some of the Portuguese took pity on him, and gave him lodgings. He was in a wretched condition when I saw him. The mate, by order of the captain, told me to advise him to make himself scarce without delay, or he would be taken on board again, and punished as a deserter. I did so, believing his sufferings, under any circumstances, could not be worse on the island than they would be if he should again be taken on board the barque. I never saw him again.

My comrade, W——, of whose sufferings during our cruise I have spoken at some length, being entirely too unwell to resume duty, was one of the number about to be left ashore. We had commenced the voyage with visionary dreams of romance and adventure. For many weeks past we had conversed together over the unfortunate step we had taken, and anxiously looked forward for a change; many weary nights had I watched by the side of my suffering friend; and, however poorly I had discharged my duty, I had the pleasure of knowing that every little attention was most gratefully felt. I was now about to part with my only friend in a foreign land, and among strangers, where a friend can best be appreciated. I need not say that the parting was a painful one. We gazed at each other with full eyes and throbbing hearts as he was about to be borne to the boat, but could not utter a word. Poor W—— had not spoken the whole morning. There was a deep touching melancholy in his looks, far more eloquent than words. All his bright hopes of recovery seemed to vanish at the thought of our separation. That I might conceal every appearance of a weakness which is looked upon by sailors as unmanly, I busied myself about the decks, knowing, too, that it was useless to repine.

Part IV—Excerpts from **Etchings of a Whaling Cruise**

At ten o'clock the order was given to "Man the waist-boat!" I was glad enough that the boat to which I belonged was chosen, as it afforded me an opportunity of going ashore. The barque lay off and on, outside the harbor. We had a hard pull against a head wind before we reached the pier, which is close by the Portuguese fort. Here we were hailed by one of the government officers, who inquired the number of sick on board, and the nature of their complaints, stating that they would not be allowed ashore if afflicted with any contagious disease. After waiting about an hour to see the American consul, Mr. Dabney, we returned to the barque, and put the invalids in the boat.

On approaching within a few hundred yards of the pier, we were hailed by a government boat bearing the national flag. It contained two or three officers, and the health doctor, a pompous and self-sufficient quack, who went through a burlesque examination of the sick men, and then gave a permit, allowing them to be carried ashore. Here they were given up to the consul, who provided them with suitable accommodations. The charges at Fayal for landing, &c., I was correctly informed, are as follows: Fee to the health doctor, four dollars; boat charges, ten dollars; for each sick man, thirty-six dollars, to be paid to the consul.

I must here mention that it was with the utmost difficulty W—— had prevailed upon the captain to let him go at all. Being part owner in the barque, he was unwilling to lose any thing in the way of fees or government charges; and ever, till we arrived in sight of the Azores, had steadily answered all W——'s petitions by the remark, *"He might rot in the forecastle!"* I did not know at this period that the captain had his eye upon a fine gold watch, which W—— had treasured for years past as a sacred token of affection from a dear relative. This watch had been committed to the captain's keeping soon after we left New Bedford. Fearing he could not get off on any other conditions, W—— offered it to him to let him go ashore. In order to keep up some show of honesty, the captain replied "that he would keep the watch, but W—— *could have it, after the voyage, by writing for it, and enclosing payment for his outfit and passage home!"*

I spent the chief part of the day in attending upon the sick. The captain procured them a passage to the United States in a small American brig bound for Bangor, Maine. While at Fayal, the captain shipped three Portuguese and two American seamen in place of the sick.

Late in the evening I bade a final good-by to my friend W——, and returned to the barque much depressed in spirits. Before daylight next

Part IV—Excerpts from *Etchings of a Whaling Cruise*

morning the light-house had faded from our sight, and, when the sun rose, it was with difficulty that we could discern on the horizon the Peak of Pico. I can not describe the feeling of utter loneliness that stole over me when once more on the bosom of the boundless ocean. Surrounded by a crew of brutal and illiterate Portuguese, I felt that I was indeed alone. When I thought of the many happy hours I had spent in W——'s society, when I looked around me, and saw objects that reminded me of him, I felt that "Othello's occupation was gone!"

The prospect before me was any thing but cheering. I dreaded to think of the long voyage, a voyage which we had scarcely yet commenced.

****

A man like our captain, whose whole soul was wrapped up in dollars and cents, could not bear with much patience a continued run of bad luck. We had killed but one whale; that disappointment alone was sufficient to render him cross-grained and ill-natured. The expense of landing the sick men was considerable; and so grievously did it prey upon his mind, that for weeks after I seldom knew him to smile. Before we had reached the Azores, he had quarreled several times with the mate. These quarrels now became more frequent and violent than ever. There was a sailor-like boldness about the brutality of the mate which the captain did not like. With the one, meanness was the prevailing trait; with the other, a devil-may-care roughness, in which he was open and above board.

One calm day a hen flew overboard. Enos, a Portuguese, was on the main-topsail yard splicing an earing. Being a very expert swimmer, and glad of the excuse to take a dive, he jumped over after the unlucky hen. We were fanning along about a knot and a half an hour. The captain, hearing our shouts of laughter as we hauled Enos in with a rope, came rushing up the companion way, roaring at the mate to "lower away a boat!" We all knew he would as soon lose his best man as a hen, and we joked Enos (loud enough to be heard aft) about being guilty of such a *fowl* piece of folly as to jump over-board after a hen. There was nothing that the captain could take hold of in this; but it irritated him. It happened that the mate was at work in the waist. Now, when captain and mate are not on the best terms, the latter generally has to bear the blame of every thing that goes wrong, and, of course, is the legitimate object of all the surplus ill humor of his sovereign master.

"Mr. D——, why didn't you lower a boat after that hen, I should think you'd have had sense enough to do that without waiting for me to tell you."

Part IV—Excerpts from Etchings of a Whaling Cruise

"I received no orders to lower a boat, sir. The man jumped overboard without asking me, and if he's fool enough to risk his life for a hen, I can't help it. You'd better talk to him about it."

"No, I'll talk to *you*!" cried the captain, very much enraged. "It was your duty to lower away a boat. Any man with an ounce of sense might know enough for that."

This of course raised the mate's "pluck" and, turning from his work, he boldly faced the captain.

"Do you suppose I'd take the responsibility to lower a boat for a cursed old hen! No, I'll be hanged if I would. You'd be the first to flare up at it yourself. Now, sir, since you've begun a jaw, I'll just tell you how we stand, Captain A——. There has been too much of this fault-finding lately. I've done my best to suit you; but, it appears to me, the more a man does to please you, the more you grumble. I've stood this long enough; so I think it's about time for us to come to an understanding about it. The amount of it is, I'll be d—d if I suffer it any longer!"

This was pretty determined language. It was such as the captain had not been used to; for, according to his own account, his former mates would lick the planks he walked on; and he had never had one to give him a back answer. He now began to draw in his horns.

"When did I find fault, Mr. D——? Tell me a single instance."

"You're always finding fault; that's enough. If we can't get along easier, the sooner we part the better. I know my place, sir, and I intend to do my duty; but I'll show you that I'm not to be brow beat and insulted!"

Some more words of a like nature passed between them, which I did not hear. There was no damage done, however. Both captain and mate remained on the worst possible terms from that time forth. They seldom spoke, except on business matters, or upon subjects connected with the voyage.

After a short cruise on the eastern ground we returned to Fayal to land another sick man. It was found necessary, when we formerly touched at the port of Orta, to ship five new hands, two of whom were Americans, two Portuguese, and an Englishman. We found them a very quarrelsome and disorderly set; but the captain had a partiality for outcast foreigners. We only remained in port a few hours, and I was not permitted to go ashore. I had the pleasure, however, of hearing that my friend W—— was rapidly improving in health. An addition to the number in the forecastle was made this time in the place of the man who was left ashore, which made the crew consist of eight Portuguese, an Englishman, and four Americans.

Part IV—Excerpts from *Etchings of a Whaling Cruise*

In the early part of our voyage we had for cook a mulatto man, who had served as a ward-room steward on board the Peacock during the United States Exploring Expedition. Whether he had acquired the habit of grumbling from his man-of-war comrades, or whether it was natural to him, I can not say; but a greater grumbler, or a more disagreeable animal, I never had the misfortune to meet. In addition to this, he had a most villainous and tyrannical temper, which continually developed itself in acts of injustice toward the crew. I had been too long living in slave states to bear very quietly the insolence of a negro, and on several occasions we came to pretty close quarters. I candidly confess, nothing but fear of the consequences prevented me from heaving the wretch overboard the first good chance. It was a source of continual annoyance to be thrown in this man's way, and particularly galling to my feelings to be compelled to live in the forecastle with a brutal negro, who, conscious that he was upon an equality with the sailors, presumed upon his equality to a degree that was insufferable. Finding I would not succumb to his insolence, as the other hands did, he took a most inveterate hatred to me, and did all in his power to render my situation unpleasant, by instilling into the minds of my comrades that I was a "broken-down dandy," who would lord it over them, if they would suffer me. As I had always made it a point never to evince the least symptom of superiority, or pretend to any thing more than those around me, he failed to effect his object in this particular; for I had the good fortune to be a general favorite. He next had recourse to another and a far more effective expedient. Our fare at the best of times was bad enough, and always scanty. When I had watch at the mast-head, or when it was my trick at the helm, he always managed to jilt me out of my allowance, or give me the offal of the crew. I had heard too many complaints made to the captain to hope for any thing from him in the way of redress. Many a night, after a hard day's work, have I turned in hungry enough to eat with relish, had it been within my reach, the common dogmeat, upon which the pampered canine gentry of the cities luxuriate. The life I had led since I had shipped produced such a change in me as made me a mere animal. When I got any thing fit to eat, which was very rarely, I devoured it with the avidity of a starving wolf. I seldom dreamed of any thing at night but good Kentucky roast beef, peaches and cream, pumpkin pies, and all the luxuries of western life.

Trifling as such things as these may appear to those who live ashore, where the poorest can by industry obtain abundance of the good things of life, they are not so trifling on board a whaler. I had seen the time when my

Part IV—Excerpts from Etchings of a Whaling Cruise

fastidious taste revolted at a piece of good wholesome bread without butter, and many a time had I lost a meal by discovering a fly on my plate. I was now glad enough to get a hard biscuit and a piece of greasy pork; and it did not at all affect my appetite to see the mangled bodies of divers well-fed cockroaches in my molasses; indeed, I sometimes thought they gave it a rich flavor.

On leaving Fayal the second time, this villainous cook, who had made such murderous attempts to starve me, was promoted steward, and a Portuguese mulatto, belonging to the Cape de Verdes, was made cook. I need scarcely say that I heartily rejoiced in the change; for I knew, let what would come, it could not be for the worse.

We were now fairly under weigh for the Indian Ocean, each day making to the southward as fast as a clumsy barque, which never sailed more than six knots an hour, except in a gale, could carry us. The monotony of a long passage is known to every body who has ever read of the sea. Seldom is it relieved, except by a squall, a calm, a sail in sight, or some trifling adventure. Time hung very heavily on our hands, though we contrived various means to pass it away as pleasantly as possible. The chief resources I had for driving dull care away were reading, drawing, writing in my journal, eating whenever I could get any thing to eat, and sleeping whenever the Portuguese would give me a chance. As to reading, I was necessarily compelled to read whatever I could get. Unfortunately, I had brought neither books nor papers with me, so that I had to depend entirely upon the officers, none of whom were troubled with a literary taste. Mr. D——, the first mate, who was very friendly toward me, had a bundle of old Philadelphia weeklies, which I read over a dozen times, advertisements and all. The cooper, a young man from New Bedford, was by far the most intelligent man aft. His stock of literature consisted of a temperance book, a few Mormon tracts, and Lady Dacre's Diary of a Chaperon. I read these till I almost had them by heart. The captain himself was an illiterate man, "wise in his own conceit." He had the reputation at home of being a *pious man*; and, as some evidence of this, I procured from one of the officers a work belonging to him of a religious character. I can not say, however, that his conduct was in strict conformity with the reputation he had gained as a man of piety. One of my shipmates had a Bible; another, the first volume of Cooper's Pilot; a third, the Songster's own Book; a fourth, the Complete Letter Writer; and a fifth claimed, as his total literary stock, a copy of the Flash newspaper, published in New York, in which he cut a conspicuous figure as the "Lady's Fancy Man." I read and re-read all these. Every week

Part IV—Excerpts from *Etchings of a Whaling Cruise*

I was obliged to commence on the stale reading, placing the latest read away till I systematically arrived at them again, when they were pretty fresh, considering the number of times they had been overhauled. When I became thoroughly satiated with the fresh and stale, I had recourse to drawing, at which I considered myself somewhat of an amateur. My stock of implements consisted of a short stump brush, a few ounces of black-lead, a piece of Indian ink, and a pen. Some of my shipmates, who had never seen any drawings in the mezzotinto style, took a great fancy to my little productions, and insisted upon having specimens for their sweethearts. By humoring them to the best of my ability, I so far gained their goodwill that they reciprocated my attempts by doing all my patching and mending, which was a very acceptable return, for I was not an expert hand at the needle. In the evening, after the decks were swept, I generally sat for an hour or two on the jib-boom playing the flute, or humming over favorite airs, many of which conjured up associations which were "pleasant, yet mournful to the soul." After one of the watches went below, we usually had a little gathering on the forecastle, and each of us told something of his past life. In this way I learned the history of all in the watch to which I belonged. Rum and love had done signal service in the way of driving them to sea.

*****

October 13th.—"There she blows!" was sung out from the mast-head.
"Where away?" demanded the captain.
"Three points off the lee bow, sir."
"Raise up your wheel. Steady!"
"Steady, sir."
"Mast-head ahoy! Do you see that whale now?"
"Ay, ay, sir! A school of sperm whales! There she blows! There she breaches!"
"Sing out! Sing out every time!"
"Ay, ay, sir! There she blows! There—there—*thar'* she blows—bowes—bo-o-o-s!"
"How far off!"
"Two miles and a half!"
"Thunder and lightning! so near! Call all hands! Clew up the fore-t'gallant-sail—there! belay! Hard down your wheel! Haul aback the main yard! Get your tubs in your boats. Bear a hand! Clear your falls! Stand by all to lower! All ready?"

Part IV—Excerpts from Etchings of a Whaling Cruise

"All ready, sir!"

"Lower away!"

Down went the boats with a splash. Each boat's crew sprang over the rail, and in an instant the larboard, starboard, and waist boats were manned. There was great rivalry in getting the start. The waist-boat got off in pretty good time; and away went all three, dashing the water high over their bows. Nothing could be more exciting than the chase. The larboard boat, commanded by the mate, and the waist-boat, by the second mate, were head and head.

"Give way, my lads, give way!" shouted P——, our headsman; "we gain on them; give way! A long, steady stroke! That's the way to tell it!"

"Ay, ay!" cried Tabor, our boat-steerer. "What d'ye say, boys. Shall we lick 'em?

"Pull! pull like vengeance!" echoed the crew; and we danced over the waves, scarcely seeming to touch them.

The chase was now truly soul-stirring. Sometimes the larboard, then the starboard, then the waist-boat took the lead. It was a severe trial of skill and muscle. After we had run two miles at this rate, the whales turned flukes, going dead to windward.

"Now for it, my lads!" cried P——. "We'll have them the next rising. Now pile it on! a long, steady pull! That's it! that's the way! Those whales belong to us. Don't give out! Half an hour more, and they're our whales!"

The other boats had veered off at either side of us, and continued the chase with renewed ardor. In about half an hour we lay on our oars to look round for the whales.

"There she blows! right ahead!" shouted Tabor, fairly dancing with delight.

"There she blows! There she blows!"

"Oh, Lord, boys, spring!" cried P——.

"Spring it is! What d'ye say, now, chummies? Shall we take those whales?"

To this general appeal every man replied by putting his weight on his oar, and exerting his utmost strength. The boat flew through the water with incredible swiftness, scarcely rising to the waves. A large bull whale lay about a quarter of a mile ahead of us, lazily rolling in the trough of the sea. The larboard and starboard boats were far to leeward of us, tugging hard to get a chance at the other whales, which were now blowing in every direction.

Part IV—Excerpts from *Etchings of a Whaling Cruise*

"Give way! give way, my hearties!" cried P——, putting his weight against the aft oar. "Do you love gin? A bottle of gin to the best man! Oh, pile it on while you have breath! pile it on!"

"On with the beef, chummies! Smash every oar! double 'em up, or break 'em!"

"Every devil's imp of you, pull! No talking; lay back to it; now or never!"

On dashed the boat, cleaving its way through the rough sea as if the briny element were blue smoke. The whale, however, turned flukes before we could reach him. When he appeared again above the surface of the water, it was evident that he had milled while down, by which maneuver he gained on us nearly a mile. The chase was now almost hopeless, as he was making to windward rapidly. A heavy, black cloud was on the horizon, portending an approaching squall, and the barque was fast fading from sight. Still we were not to be baffled by discouraging circumstances of this kind, and we braced our sinews for a grand and final effort.

"Never give up, my lads!" said the headsman, in a cheering voice. "Mark my words, we'll have that whale yet. Only think he's ours, and there's no mistake about it, he will be ours. Now for a hard, steady pull! Give way!"

"Give way, sir! Give way, all!"

"There she blows! Oh, pull, my lively lads! Only a mile off! There she blows!"

The wind had by this time increased almost to a gale, and the heavy black clouds were scattering over us far and wide. Part of the squall had passed off to leeward, and entirely concealed the barque. Our situation was rather unpleasant: in a rough sea, the other boats out of sight, and each moment the wind increasing.

We continued to strain every muscle till we were hard upon the whale. Tabor sprang to the bow, and stood by with the harpoon.

"Softly, softly, my lads," said the headsman.

"Ay, ay, sir!"

"Hush-h-h! softly. Now's your time, Tabor!"

Tabor let fly the harpoon, and buried the iron.

"Give him another!"

"Ay, ay! Stern all!"

"Stern all!" thundered P——.

"Stern all!"

And, as we rapidly backed from the whale, he flung his tremendous

Part IV—Excerpts from Etchings of a Whaling Cruise

flukes high in the air, covering us with a cloud of spray. He then sounded, making the line whiz as it passed through the chocks. When he rose to the surface again, we hauled up, and the second mate stood ready in the bow to dispatch him with lances.

"*Spouting blood!*" said Tabor. "*He's a dead whale!* He won't need much lancing." It was true enough; for, before the officer could get within dart of him, he commenced his dying struggles. The sea was crimsoned with his blood. By the time we had reached him, he was belly up. We lay upon our oars a moment to witness his last throes, and, when he had turned his head toward the sun, a loud, simultaneous cheer burst from every lip.

A low, rumbling sound, like the roar of a distant waterfall, now reached our ears. Each moment it grew louder. The whole expansive arch of the heavens became dark with clouds tossing, flying, swelling, and whirling over and over, like the surges of an angry sea. A white cloud, gleaming against the black mass behind it, came sweeping toward us, stretching forth its long, white arms, as if to grasp us in its fatal embrace. Louder and still louder it growled; yet the air was still and heavy around us. Now the white cloud spread, whirled over, and lost its hoary head; now it wore the mane and fore feet of a lion; now the heads of a dragon, with their tremendous jaws extended. Writhing, hissing, roaring, it swept toward us. The demon of wrath could not have assumed a more frightful form. The whole face of the ocean was hidden in utter darkness, save within a circle of a few hundred yards. Our little boat floated on a sea almost unruffled by a breath of wind. The heavy swell rolled lazily past us; yet a death-like calmness reigned in the air. Beyond the circle all was strife; within, all peace. We gazed anxiously in each other's faces; but not a word was spoken. Even the veteran harpooner looked upon the clouds with a face of unusual solemnity, as we lay upon our oars, awed to silence by the sublimity of the scene. The ominous stillness of every thing within the circle became painful. For many long minutes the surface of the water remained nearly smooth. We dreaded, but longed for a change. This state of suspense was growing intolerable. I could hear the deep, long-drawn respirations of those around me; I saw the quick, anxious glances they turned to windward; and I almost fancied I could read every thought that passed within their breasts. Suddenly a white streak of foam appeared within a hundred yards. Scarcely had we unshipped our oars, when the squall burst upon us with a stunning violence. The weather side of the boat was raised high out of the water, and the rushing foam dashed over the gunwale in torrents. We soon trimmed her, how-

Part IV—Excerpts from *Etchings of a Whaling Cruise*

ever, and, by hard bailing, got her clear of water. It is utterly impossible to conceive the violence of the wind. Small as the surface exposed to the squall was, we flew through the foaming seas, dragging the dead body of the whale after us with incredible velocity. Thus situated, entirely at the mercy of the wind and sea, we continued every moment to increase our distance from the barque. When the squall abated, we came to under the lee of the whale, and looked to leeward for the barque. Not a speck could be seen on the horizon! Night was rapidly approaching, and we were alone upon the broad, angry ocean!

"Ship your oars," said the headsman; "we'll not part company with old Blubber yet. If we can't make the barque, we can make land somewhere."

"Ay, ay," said Tabor, with a sly leer, "and live on roast-beef and turkey while we're making it."

With heavy hearts and many misgivings we shipped our oars, heartily wishing the whale in the devil's try-pots; for we thought it rather hard that our lives should be risked for a few barrels of oil. For two hours we pulled a long, lazy, dogged stroke, without a sign of relief. At last Tabor stood up on the bow to look out, and we lay on our oars.

"Well, Tabor, what d'ye see?" was the general inquiry.

"Why," said Tabor, coolly rolling the quid from his weather to his lee cheek, "I see a cussed old barque that looks like Granny Howland's washtub, with a few broom-sticks rigged up in the middle of it."

"Pull, you devils!" cried P——; "there's duff in the cook's coppers."

"Yes! I think I smell it." said Tabor.

It was nearly dark when we arrived alongside of the barque with our prize; but what was our surprise to find that the starboard and larboard boats had killed *five* whales between them! They were all of a small size, and did not average more than fifteen barrels each.

That night not a breath of air ruffled the clear, broad ocean as it swelled beneath and around us, forming a multitude of mirrors that reflected all the beauties of the splendid canopy above. The moon arose with unusual brilliancy. It was a night for the winged spirits of the air. I enjoyed a melancholy pleasure in walking the decks beneath the soft moonbeams, thinking of past times. Silence reigned over the deep. The calm, broad ocean presented a beautiful simile of repose, and the light, shadowy clouds floated motionless in the air, as if in awe of the mighty wilderness of waters beneath them. A clear, dreamy haze upon the horizon. I gazed with pensive feelings upon this scene; so calm, so heavenly, so unrivalled

Part IV—Excerpts from Etchings of a Whaling Cruise

in its loveliness; and I thought, with a sigh, of the coming day : the fiery, tropical sun; the loud, harsh voices of the officers giving orders; the heat and smoke of the try-works; and all the realities of a whaleman's life. I have heard of the solitude of the desert; but what can compare with that of the ocean at such a time as this.

Never had the sea looked more beautiful than it did that night. It was a source of pleasure to feel that, notwithstanding the wretched life I led, there were still left a few of the better feelings of my nature. A passage in the "Vision of Don Roderic" occurred to me as singularly expressive of the checkered fortunes of a sea-farer. Well might I hope the light cloud which occasionally obscured the moon's brightness might prove a happy omen of my future fate:

> "Melting, as a wreath of snow it hangs
> In folds of wavy silver round, and clothes
> The orb in richer beauties than her own;
> Then, passing, leaves her in her light serene."

At daylight next morning all hands were called, and set to work getting up the cutting tackle, and making other preparations for cutting in. As this process of "cutting in" seems to be but imperfectly understood by those who have not been engaged in the business, it would perhaps be well enough to devote a page or two of description to it in this place.

When the whale has been towed alongside by the boats, it is firmly secured by a large rope attached to the "small" by a running noose. There is not a little ingenuity in the manner in which the fluke rope is first passed under the body of the whale. A small line, to which a lead is fastened, with a block of wood at the extremity, several fathoms from the lead, is thrown over between the whale and the ship's side. From the impetus given to the lead, it sinks in a diagonal direction, drawing the block down after it. One end of the lead line is fastened to the end of the fluke rope on board, and the block attached to the other rises at the off side of the whale. It is then hauled on board by means of a wire hook fastened to a long pole, and, in hauling it in, the fluke rope passes round under the body of the whale, till the end arrives on board, when it is passed through the loop in the other extremity, and thus a running noose is formed, which is easily slipped down to the small. The fluke rope is then made fast on the forecastle, and the flukes are hauled up to the bow, or as near as they will reach, leaving the head pointed aft. Of course, the size of the vessel and the length of the whale make a great difference; but in general the head reaches to the quarter. To prevent concussion, the whale is always on the weather side. The progress of the vessel, which is usually under easy sail during the time of cutting in, keeps the whale from drifting out at right angles from the side; though, in most cases, the head is kept in its appropriate position by a small rope made fast aft.

The cutting tackle is attached to a powerful strap, or pendant, passing round the mast in the main-top by two large blocks. There are, in fact, two tackles, the falls of which pass round the windlass. To each of these tackles is attached a large blubber hook, which, upon being made fast to the blubber, are hauled up by the windlass, one only being in operation at a time, so that when the first strip of blubber, or "blanket piece" reaches the stationary block on the pendant, the other can be made fast by a strap and bolt of wood to a hole cut below the point at which that blanket piece is to be cut off. I have endeavored to give some idea of this part of the process in the frontispiece accompanying the work. The blanket pieces are stripped off in a spiral direction, running down toward the flukes; the whale turning, at every heave of the windlass, till the whole covering of blubber is stripped off to the flukes, which are hoisted on board, and those parts containing oil cut away, and the remainder thrown overboard. The head having, in the first place, been cut off and secured to the stern, is now hauled up, with the nose down, if too large to be taken on board, and hoisted as far out of the water as may be found convenient, and the oil or liquid spermaceti bailed out with a vessel attached to a long pole, and thus taken in and saved. As there is no little risk attending this mode of getting the spermaceti, and a great deal of waste, the head is always taken on board, when not too large or heavy.

The "case," which is the name given by whalers to the head, sometimes contains from ten to fifteen barrels of oil and spermaceti. A single "blanket piece" not unfrequently weighs a ton or upward. In hauling it up by the tackles, it careens the vessel over frequently to an angle of fifteen or twenty degrees, owing to its own great weight, combined with that of the whale, the upper surface of which it raises several feet out of the water. When the blanket piece has reached the stationary block in the top, it is cut off by a boat-steerer, who stands by with a boarding knife, having first, however, been secured below by the other blubber hook, which is hauled taught, to prevent it from breaking away by too sudden a jerk. The upper piece then swings in, and, when it ceases its pendulating motion, is dropped down into the hold or blubber-room, where it is cut up into blocks of a foot and

Part IV—Excerpts from Etchings of a Whaling Cruise

a half or two feet in length, and eight or ten inches in width. These blocks are called "horse pieces." The white, hard blocks, containing but little oil, and which are found near the small, and at the flukes, are called "white horse." The carcass of the whale, when stripped of its blubber, is cast loose, and soon sinks from the want of its buoyant covering. I have seen it float astern, however, some distance without sinking.

Breakfast over, all hands were called to cut in. Six or eight men were stationed at the windlass, two in the blubber-room, and the boat-steerers in the waist. The first and second mates took their station on a couple of stages, or platforms, rigged out at the gangway, each provided with a spade. One of the boat-steerers, whose turn it was to fasten the blubber hook, went down over the side on the whale's back, and, after several unsuccessful attempts, and rather an uncomfortable ducking, performed his task. While yet on the whale's back, a large, hungry-looking shark, which had been eyeing him for some time very anxiously, was washed up behind him by a heavy sea, and apparently loath to lose so good an opportunity of making a meal, began to work his way along the slimy surface till within a foot or two of the boat-steerer's heels. The officers happened at the moment to be looking up at the pendant block, and in all probability the man would have been seriously injured, if not carried off bodily, but for the timely alarm of one of the crew. The mate immediately turned to see what was the matter, and perceiving the critical position of the boat-steerer, brought his spade to bear upon the shark, and at a single dart chopped off his tail. Strange to say, the greedy monster did not appear to be particularly concerned at this indignity, but, sliding back into his native element, very leisurely swam off, to the great apparent amusement of his comrades, who pursued him with every variety of gyrations. It surprised me to see with what cool indifference the boat-steerer witnessed the whole transaction. I do not remember that he said a word about it.

The various duties being apportioned to the men without favor or choice, it fell to my lot to sit on the weather side of the quarter-deck and turn the grindstone; a tiresome and monotonous task. The cooper attended to the sharpening of the spades, boarding knives, and other implements used in "cutting in." I am not sure that I had the hardest of the work to do, but it certainly was the most unpleasant; for I could not prevail upon any of the hands to change places with me, even for a brief period. My appearance at this time would have been somewhat striking to some of my friends in Washington. With my duck frock all black with whale-gurry, my trowsers torn and smeared with rough work, my red Scotch cap half-way over my eyes, and my face oily and sunburned, I certainly looked as little like my original self as one can well imagine. There I turned that grindstone, and turned on hour after hour, and turned the palm of my right hand into a great blister, and turned the palm of my left into another; turned both my arms into a personified pain; turned every remnant of romance out of my head; turned and turned till my grand tour seemed to have turned into grindstone; round and round I turned that stone till I began to think I was a piece of the handle, and turned with it; and my head appeared to turn, and my feet to turn, and the game-legged cooper to turn, and the ship to turn, and the sea, and the whale, and the sharks, and the clouds, and all creation seemed to be turning with myself and that grindstone! Having at last contrived to get a sufficient number of spades sharpened ahead of the mates, I peeped over the quarter-rail to see how they were getting on. The sharks had by this time gathered around the vessel in immense numbers, and eight or ten were fighting just under the quarter for a piece of the whale's carcass which had been cut away. Watching my opportunity, I snatched up a spade, slipped it over while the captain was forward, and began a terrible onslaught among the sharks. With five or six thrusts I killed four of the greedy monsters, by striking them on the back of the head, and cutting the principal artery. This was quite a refreshing little episode in my business of turning; and my success in the destruction of sharks induced me to believe that I had a greater natural *turn* for sport than the monotonous *turning* of a grindstone. But my amusement was of short duration. The eagle eye of the captain espied me before I could get in the long pole of the spade.

"Ha! what are you at there?" cried his highness directly behind me, at the very moment when I supposed he was on the forecastle giving orders to the men. "What are you at, hey?"

"Keeping off the sharks, sir."

"Who told you to keep 'em off?"

"Nobody, sir."

"Haul in that spade directly!"

"Ay, ay, sir!"

"And, look'ee, if I catch you keepin' off any more sharks, I'll wipe you down with a rope's end!"

"Ay, ay, sir!"

"Cooper, hain't you got no work for this fellow?"

Part IV—Excerpts from Etchings of a Whaling Cruise

"Not just now, sir."

"Go to the windlass, then, and rest yourself on a handspike!"

"Ay, ay, sir!"

I had reason to consider my shark-killing a poor speculation. The heaving and surging at the windlass was but a questionable improvement upon my old business of turning the grindstone. At the word, "Heave away!" somebody struck up an extemporaneous song, which, to the best of my recollection, had no particular claims to poetical merit, but ran somewhat thus:

> "Heave him up! O he yo!
> > Butter and cheese for breakfast
> Raise the dead! O he yo!
> > The steward he's a makin' swankey.
> Heave away! O he yo!
> > Duff for dinner! Duff for dinner!
> Now I see it! O he yo!
> > Hurrah for the Cape Cod gals!
> Now I don't. O he yo!
> > Round the corner, Sally!
> Up she comes! O he yo!
> > Slap-jacks for supper!
> Re—re—ra—ra—oo—we ye yo ho! Them's 'um!"

At the conclusion of this medley, the captain, who had seated himself in the starboard quarter boat to inspect the cutting, began to criticise the mate's style of cutting rather severely. Now the mate, be it known, was really a very skillful whaleman, and handled the spade with an unerring hand. The "old man's" comments, thus lowering him in the eyes of the crew, by no means pleased him.

"I say, Mr. D——," persisted the captain, "that's not the way to cut in a whale. I don't want no such work as that about me."

"It's my way, sir," replied the mate, getting very red in the face.

"Well, I never see a whale cut that way. I ain't used to it; I won't have it."

"You haven't seen every thing yet, sir. I've always cut whales this way, and always mean to do it."

"No you won't; not here you won't. You can just cut as I tell you."

"I reckon I know my own business, Captain A——. Now, sir, I'm not a going to be dictated to in this manner. If you think you can cut a whale better than I can, you'd better take my place."

Part IV—Excerpts from *Etchings of a Whaling Cruise*

After some more angry words, during which both the captain and mate became much excited, and threatened to whip each other, the quarrel ceased, and the "old man" went below in high dudgeon.

This was all "nuts" for me. I was rejoiced to see somebody among the privileged few talk up to him as he deserved. Indeed, I was itching to express my own personal opinion on the subject, but had no particular fancy for the mode of "wiping down" hinted at a short time previously.

Cutting in, trying out, and clearing up the decks, occupied us for the next six days. We had an average of five hours' sleep out of the twenty-four. Working incessantly in oil, which penetrated to the skin, and kept us in a most uncomfortable condition, besides being continually saturated with salt water, produced a very disagreeable effect upon those who were not accustomed to such things, by chafing the skin, and causing painful tumors to break out over the whole body. Before I had half finished my share of the labor, I heartily wished myself in the meanest dog-kennel ashore, or, to borrow an old idea, I should have considered myself in an enviable situation had some enemy been kicking me down Pennsylvania Avenue. Tabor, the oldest whaleman on board, who laughed at hardships, and took all the disagreeable parts of his duty with the utmost good humor, frequently joked me on my "grand tour to Europe."

"What do you think of whaling now, B——? Is it equal to traveling in Italy?"

"I think not, Tabor."

"Tain't writing short-hand neither, is it? I think you'd as lief be in Washington, with them thar big members of Congress, as blubber-hunting. Cutting figures with the pen ain't cutting blubber, by a considerable sight, is it?"

I freely acknowledged that, of the two sorts of cutting, I preferred cutting figures with the pen; at which Tabor laughed most heartily, assuring me that "it was nothing when I'd get used to it. By'm-by I'd see what whaling was. This wasn't a circumstance. I'd smell smoke yet. I'd begin to find out what some folks was at while others was riding about in chaises."

We had an extra supply of meat on this occasion, with about a quarter of a pound of rancid butter, which was to be divided among all hands. This unusual liberality on the part of the captain astonished us all, and filled our hearts with gratitude. We took the saucer containing the precious morsel, and, seating ourselves in a circle round it, enjoyed our good fortune by various amusing comments upon the captain's unparalleled liberality.

Part IV—Excerpts from Etchings of a Whaling Cruise

Of all things in the world, sailors despise most a stingy, thin-skinned captain. They will excuse cruelty, unnecessary hardship, or coarse and brutal language, for they become accustomed to it; but any thing like stinginess or meanness they heartily detest. Bill Mann growled like a sick bear, protesting, in his own peculiar style, that it was "the blamdest thing he ever saw done aboard any ship. He wished his soul might everlastin'ly stick fast in purgatory, if he wouldn't tell the *counsel* of it." Mack wanted to carry it back to him with the thanks of the crew, "hopin' he wouldn't rob himself, for the men was afraid to eat it, bein' as they had never seen anything like it since they'd follered the sea." This proposition was negatived, and we at last agreed to mix the butter up with a pan of dirty bread and heave it overboard. The captain, who was sitting in the stern boat, chanced to spy the bread as it floated toward him, and seizing a pole with a small net attached to it, which he always kept in the boat, he hooked up every morsel of it. Owing to the round-house concealing him from our view, we knew nothing about this, till he came forward about fifteen minutes after, with a plate in his hand, containing what we supposed to be an additional treat for all hands. There was a fiendish smile of triumph visible about his lips, however, and a ferocity in his eyes that boded us no good. Holding the plate out toward us, he pointed with his forefinger at the startling apparition of the resuscitated bread, and demanded, in a deep, distinct voice,

"Which on ye did that, hey? Look at it, every one o' ye; examine it well. Did ye never see it before, hey? Taste it; it's got salt water in it, but it's good, hey? A nice set of darned rascals! Don't get enough to eat, hey? I starve you, do I, hey? You don't like butter; oh no, you can't eat it! Nice stomachs, I'll swan! Whose work is this? Don't all answer at once! Who did it?"

No one answered. We all felt that we were guilty, and it is not to be wondered at that, taken aback so suddenly, we could account for the unexpected reappearance of the bread and butter, which we had supposed was food for the sharks, in no other way than by presuming old Skinflint was in league with the devil.

"Oh, you didn't do it, none on ye!" shouted the captain, letting loose his wrath. "It grew in salt water! It wasn't hove overboard at all! Well! take and eat it now; and, mark my words, the first man I catch heavin' good vittals overboard, I'll heave him overboard!" With that he flung it down before us and walked aft, grieving over our depravity and his probable loss. From that time forth he used to sit in the stern boat for hours every day, dodging

Part IV—Excerpts from *Etchings of a Whaling Cruise*

his net in the water at every thing that looked like biscuit or meat. Sometimes he would catch up what he didn't bargain for, and his low, half-smothered comments, audible only to the man at the wheel, would afford us infinite amusement. Patience and perseverance finally rewarded him with success. He had been at his post regularly three times every day for about a month, fishing up whatever attracted his insatiate eye, when one day he made a haul of a fine fat piece of pork. He jerked it in, chuckling over his good fortune, and muttering, "Aha! I've caught you at last, you infernal scoundrels! I'll give you fat pork to throw away in a hurry again!" Calling to the steward to pass up a fork, he spitted it handsomely, and carried it forward for our inspection. Looking each of us hard in the face by turns, he demanded, in a voice of thunder, "Who hove the pork overboard?"

No answer.

"Won't you tell me, you sheepheads?"

No answer yet.

"You won't, eh? It's your work, M'F——?"

"No, sir, tain't. Pork's scarce about these diggins. I don't throw away a good chunk when I get hold on't."

"It's you then, Vernon. I'll skin you for it. I'll show you how to waste good meat, you worthless bullet-head. You don't earn your salt."

"Twasn't I, sir. I was down in the forecastle."

"Then you know who did it?"

"No, sir, I don't."

"What the devil *do* you know?"

"I know I hadn't enough o' meat for dinner."

"You hadn't, eh? Well, *I'll* see to it. You're a parcel of hogs, that's what you are! Cook, from this time forth cut these men's meat up in small pieces, and just give 'em as many pieces as'll go round."

"Ay, ay, sir."

"Now I'll know when meat's wasted again, and why." So saying, he walked aft, satisfied at least that half a pound lost was a barrel gained. Whether the piece which he picked up had been thrown over purposely, or had fallen from the top of the caboose accidentally, I never could find out; but this much I know, our share of meat soon resembled the Irishman's dinner of "potatoes and point" barrin' the potatoes.

It is customary, in most vessels, to give the watches their meals alternately, the watch below being served first. This prevents confusion and quarreling. The watch on deck, when relieved, can then enjoy their meals

Part IV—Excerpts from Etchings of a Whaling Cruise

at their leisure. Sailors generally, though any thing but deficient in appetite, have great respect for the decencies of civilized life on such occasions as these. I have never seen selfishness or greediness on board well-disciplined ships during meal times. On the contrary, it is surprising to find a very delicate sense of propriety among men who have had so few opportunities of cultivating the refinements of social intercourse. I have often seen a well-behaved and orderly crew seated around the forecastle, eating their meals in peace and good-humor, and each one neglecting no opportunity of extending a courtesy to his shipmate. This is almost invariably the case where they are well treated by the captain and officers. Like children, they can be moulded to almost any thing; and where a bad example is set aft, the best of them will be sure to follow it. No one who has never been to sea and witnessed such scenes, can conceive the importance of these little forms of politeness, and their happy effects upon the crew. Where there is a kindly feeling on the part of every man toward his neighbors, the worst fare can be eaten with relish.

How different it was with us! We had been so accustomed to see quarreling between the captain and his officers, and so much discontent manifested by the latter about their meals, that it would have been a miracle if we had not imbibed the prevailing spirit. But we had our frailties too, and were not proof against the example of these high authorities. There was some excuse for us, however; we really had something to be discontented about. The captain had shipped a gang of voracious and filthy Portuguese, whose condition had never been better than that of swine, and with these uncultivated brutes we were compelled in self-defense to do the best we could for ourselves. It was degrading to the pride of those who were burdened with that inconvenient commodity to be obliged to rush like hungry wolves for a mouthful of meat at meal-times; but there was no help for it. We either had to join in the struggle, or lose our allowance; for it was seldom there was enough before us to satisfy half our number.

Although it was no joke to be starved, I always found food for merriment when I could get nothing of a more substantial nature. The cook, in order to see fair play, generally watched his opportunity, and, when the hands were scattered around the forecastle, he would pitch the meat kid down on the deck, and sing out, at the top of his voice, "Meat! meat! fall to, all hands!" This startling intelligence never required repetition. Those who were nearest would jump up and run toward the smoking morsel as if simultaneously stung by a score of wasps. Those who, unfortunately, happened to be at a distance, had no resource but to dart after their leaders in the rush, and, by dint of hard struggling, secure a place by the meat kid. There was something indescribably ludicrous in the earnestness with which we all entered into the contest. It was not exactly a struggle involving "life or death," but it was of scarcely less importance; for "Meat, or no meat?" was the grand question. Nor did we hesitate to resort to the most cunning expedients to obtain our fair proportion of the salt junk. When hard pressed, it was not unusual to pick up a rival and carry him back ten or twelve paces, and, before he could regain his legs, take advantage of his absence, and get in six feet ahead of him. Some had been shouldered away so often in this manner by those who were larger and stouter, that hunger taught them a new expedient; and they secured their share on several occasions by working in like eels under the legs of those who were ahead of them. Big John, the Portuguese, having the advantage of us all in size and strength, would sometimes make a clean sweep with his arms, and lay half a dozen of us sprawling on the deck; but, being less greedy than the rest of the Portuguese, he never took more than his share, and only exerted his powers in this way for the sake of amusement.

****

October 26th.—Made the Isle of Sal, one of the Cape de Verdes, distant thirty miles. The peak is of very considerable height, and bears some resemblance to the Peak of Pico. We ran down along the shore, which has a beautiful and fertile appearance, and had a fine view of the principal harbor and town. There were several vessels in port, taking in cargoes of salt for South America. Lay to all night, and next morning at daylight made sail for Bonavista. Learning there that there was an American vessel in port at the Isle of May, we hauled off and ran down for that island, the captain being desirous to send home the oil we had on board.

A little incident occurred during the day which afforded us all much amusement. D——, the cooper, was in the habit of "raising whales" when nobody else on board could see them; and as there was a bounty up for the first whale yielding fifty barrels, he was always on the alert. While we were standing by the braces, waiting for orders, we heard him singing out from the mast-head with all his might,

"Thar' she breaches! Thar' she breaches!"

"Where away?" said the captain.

"A point off the weather bow, sir. Thar' she breaches!"

Part IV—*Excerpts from* Etchings of a Whaling Cruise

"How far off?"

"Ten miles. Thar' she breaches! Thar' she breaches!"

"Don't she blow at all?"

"No, sir; there's no spout; nothing but breaches—very large breaches. *Thar'* she breaches!"

"Luff up to the wind. Do you hear, at the wheel? Cooper, are those breaches in sight now?"

"Yes, sir; I see them all the time. She don't stop breaching at all—large breaches! It must be a very large sperm whale. Thar' she breaches! Thar' she breaches!"

"What the deuse! don't the breaches stop at all?"

"No, sir. Thar' she breaches! *thar*.—"

"Sing out every time. Get your boats ready, and call all hands."

"Thar'—thar'—*thar'* she breaches!" shouted the cooper, from the mast-head. For upward of fifteen minutes he strained his lungs in this way, when he suddenly became silent.

"Where's the whale now?" said the captain.

"I don't believe it's a whale, sir," replied the cooper, in a tone of disappointment.

"What in the nation do you call it, then?"

"Why, I don't know exactly. It looked very much like a whale at first."

"How now? Don't you know a whale when you see one? What is it?"

"Well, I don't know, sir. It ain't a whale, that's certain."

"You don't know, you infernal sheephead! Steward, pass up my spyglass!" and, taking a steady look from the main-top in the direction indicated by the cooper, he suddenly exclaimed, "Why h—ll and d—n—n! *that's Leton's Rock*!"

We all enjoyed a hearty laugh at the cooper for his mistake. The old Portuguese cook, who was something of a wag, rallied him most unmercifully. For weeks after, when the hands would gather in the waist for a dance, old Slush, grinning from ear to ear, would gaze toward the horizon with eyes like saucers. This was a signal for some of the crew to sing out, "Halloo, Slush, what d'ye see?"

"Large sperm whale, sare! Dar' she breach! Ten miles off, sare! Dar' she breach! She breach all de time, sare! Dar' she breach! Big whale, sare—dat big whale! He, he, he! yaw, yaw, yaw! *Dar'* she breach! Cooper, you sabe big rock."

In spite of the bursts of laughter which invariably followed this sally of wit, the cooper maintained the utmost good-humor, and always joined in the fun.

****

It is usual in whalers to get up a bounty, by way of encouragement to the look-outs aloft. This bounty is sometimes at the expense of the owners, who offer it with a view to promote vigilance on the part of the crew, that the voyage may be as short and profitable as possible. Five or ten dollars reward for a whale to be "raised" by any given time sharpens the sight of the men at the mast-head amazingly. Whalemen, however prodigal of their earnings ashore, are very different at sea. The desire to make a good voyage seems to be the mainspring of all their actions. With what reckless liberality the proceeds of their industry are spent when they arrive in port I need not say, for the open-hearted character of Jack ashore is known all over the world. From the close calculations which they make at sea, one would think they were the most penurious race of men in existence; but such is far from being the case.

In lieu of any bounty from the captain or owners, the crew frequently get up a system of reward on their own account. This plan is often followed by the best results. It inspires a spirit of emulation among them that gives rise to great activity and vigilance.

I present as a specimen a copy of a paper signed by the crew:

"The undersigned, hands before the mast, agree to pay the sum affixed to our respective names, on every barrel of oil raised by a subscriber, to said subscriber; the oil to be measured as stowed down."

This may require a few explanatory remarks. In the first place, to "raise oil" is an expression peculiar to whalemen. The man at the mast-head, who is the first to discover a whale, "raises oil" provided the whale be taken. If a subscriber raise a hundred barrels of oil, according to the agreement (two cents a barrel being the sum affixed to each signature), he is entitled to two dollars from each of the other subscribers, which, allowing that there are ten subscribers besides himself, makes twenty dollars. By this arrangement he may earn sufficient for spending money during the voyage. The chances are equal. The most vigilant subscriber makes the most money, and the most careless loses most. It is customary to make the sums affixed to each signature proportionate to the lay of the subscriber: a green hand paying a cent on every barrel, an ordinary seaman a cent and a half, and an able seaman two cents, or whatever rate may be agreed upon. Those who do not

Part IV—Excerpts from Etchings of a Whaling Cruise

choose to subscribe have, of course, nothing to do with it; but it is generally the case that this class is composed of the most worthless of the crew. Consequently, there is a constant competition among the vigilant portion of the crew; and if there is little success, it is not owing to carelessness or neglect of duty on their part. Those who are desirous of promotion can not better evince their claims than by activity and vigilance in this branch of the business; for as it is a primary object in whaling to see whales when they appear above the surface of the water, so it is the chief qualification of a good whaleman to understand thoroughly the different species of whales, and how to distinguish them.

November 2nd.—Since we entered the tropics I have frequently enjoyed the beauties of a sunrise at sea, which I think are more gorgeous in these latitudes than farther north or south. I never saw anything to compare with the splendors of the scene which I witnessed this morning. We appeared to float in an immense arena, encircled by ranges of hills of the most magnificent and brilliant colors. The sea was perfectly calm; and as the sun burst through from the east, gilding the edges of this mighty inclosure with the richest hues, such a combination of lights and shades was visible as to form a world of visionary splendor rather than any thing earthly: the clouds ever changing into the most fantastic and beautiful forms; sometimes assuming the appearance of a group of fairy islands, resplendent with cities and palaces of gold, and at others bearing a strong resemblance to a bold, rugged chain of mountains capped with snow, glancing brilliantly in the sunbeams. But such sights as this are not to be described; they must be seen.

****

The last subject to which I shall here allude is that of ship-keeping. In whalers there is a shipkeeper, or a man who attends to the ship when the boats are lowered. He is either chosen from among the foremast hands, or shipped at the port from which the vessel sails. The duty of the shipkeeper is by no means unimportant. The safety of the boats frequently depends upon his vigilance and knowledge of sea-craft; and, in cases of accident, the lives of the boat's crew are often dependent upon him. It is the duty of a ship-keeper to keep the run of the whales when the boats are lowered, and to make the various signals necessary to indicate their situation to the boats. Our signals were arranged in the following order:

Part IV—Excerpts from *Etchings of a Whaling Cruise*

Whales up. Signal at the main'top-gallant-mast.
Whales on the weather bow. Weather clew of the fore-top-gallant-sail or fore-top-sail up.
Whales on the lee bow. Lee clew up.
Whales on the weather beam. Weather clew of the fore-top-gallant-sail or fore-top-sail up, and waif pointed to windward.
Whales on the lee beam. Lee clew and waif.
Whales ahead. Jib down.
Whales between the boats and ship. Colors at the fore and main'top-gallant-masts.
Boat stove. Colors at the fore and mizzen.
Come aboard. Colors at the peak.

In small vessels the ship-keeper is allowed two hands to assist him in working the ship; but the number depends more upon the state of the weather than the size of the vessel.

****

December 21st.—Just as the larboard watch was called (at twelve M.), the man at the mast-head sung out, "There she blows!" I had turned out, and was about to go on deck, when I heard the word given to lower away the mate's boat. During our watch below, the waist boat had lowered after a school of finback whales. She was now about five miles off, in the midst of the school. The watch on deck manned the larboard boat, leaving six or eight hands to take care of the ship. Scarcely had she touched the water, when the whale rose within a few yards of our lee bow. It was perfectly calm; the surface of the water was of glassy smoothness. The whale was distinctly visible as he rose to blow.

"That's a right whale!" said the captain, who had ascended the mizzen shrouds to watch the movements of the boats. "Give him a dart! Don't stave your boat!"

The boat was close upon him in an instant. Hitherto he seemed unconscious of the noise and confusion around him, or of his proximity to the ship. As the words were echoed back from the boat, and the splash of the oars, as she backed away, fell upon his ear, he seemed to be a little alarmed, and turned flukes, going lazily downward in a diagonal direction. Antone, the boat-steerer, let fly his iron, but the distance was too great, and it took no effect.

Part IV—Excerpts from Etchings of a Whaling Cruise

"Oh the devil!" shouted the captain, in a tone of disappointment; "I'd have given five dollars for that chance. You ain't worth your salt, you twopence head!"

While Antone was hauling in the iron, and grumbling at his bad luck, the whale took a little circuit. He was visible at a great depth through the transparent water. The man at the mast-head continually indicated his position, being enabled to see him at the depth of fifteen or twenty fathoms. He rose, at length, within a few feet of the waist, and commenced blowing.

"Pull ahead! pull all! now's your time!" cried the mate.

"Pull all!" was echoed back by the crew, and the boat was within dart of him in three or four strokes. Antone was so eager to make up for his first failure that he overshot the mark this time. The iron slightly pricked the whale. Plunging down again, the huge creature milled round the stern. The boat followed close in his wake; but his evolutions were so sudden that it was difficult to get within dart of him. In about ten minutes he rose directly under the jib-boom. Antone stood ready with his iron. Watching his opportunity, he darted as soon as the boat came bow across the head. A tremendous hollow roar, like that of an infuriated bull, issued from the wounded monster. The blood spouted in torrents from his wound. Lashing the water with his flukes, he plunged down, covering the boat's crew with clouds of blood and spray.

"You've killed him! that whale's a fool!" cried the captain.

"He's dead! he's dead!" shouted Antone, greatly excited; "I've fixed him!"

"Dead be d—d! Clear your line!" thundered the mate. "Hold fast now! Pull two oars! back three! Pull all, now! Mind what you're about there, Mack. D'ye want to get stove? Take a reef in your eyes, and keep 'em aft here."

The moment the line was made fast, the boat dashed right under our stern with fearful velocity. The whale sheered off barely in time to avoid dashing the boat to atoms against the lee quarter. As he rose within a few fathoms of the ship, he uttered another frightful roar, and the blood streamed from him in torrents, discoloring the water entirely around the vessel, so that she actually appeared to float in a gory sea. It was evident, however, that no vital part had yet been touched. The mate now sprang to the bow of the boat, shouting, "Pull, my lads, pull!" Before the crew could stop her headway, the whale's head rose about six feet out of the water,

Part IV—Excerpts from *Etchings of a Whaling Cruise*

within half a dart of the boat. Shooting out his lance, the mate gave him a gentle prick on the nose; and dashing down with a hollow groan, the goaded monster made straight for the ship, towing the boat, with incredible swiftness, toward the weather beam. For a moment I thought nothing could save her. Her bow was not more than six feet from the vessel, when the whale suddenly milled, and thus saved her, and perhaps the lives of some of the crew.

"Stand by to lower the starboard boat!" cried the captain. "*I'll* have that whale. *I'll* see whether he can be killed or not. *I'll* not lose a good chance. He won't have a fool to deal with if *I* get within dart of him. Stand by all! Man the dart tackles, and lower away!"

A moment more, and the starboard boat was in hot pursuit. Bill Mann and I were left to take care of the ship this time: a circumstance which I did not regret, as the sun was pouring down with a burning intensity. I went to the mast-head, that I might enjoy a better view of the chase. It was now truly exciting. We hoisted a signal for the waist boat, then about three miles from the scene of action. The mate's boat ploughed the water at the rate of ten knots an hour, and increased in speed as pain gave fresh impulse to the whale. In about an hour the three boats were in a line, running to the leeward at a brisk rate. The larboard boat were as head-and-head with the whale at his next rising, and the waist boat rapidly bearing down upon him in an opposite direction. Ere the lances of either could be made use of, he rolled over in his agony, and parted the iron of the fast boat with a furious struggle. He then sounded, leaving the three boats in a whirlpool of blood and foam. When next seen, he was spouting blood a mile off; but it was so late in the evening that the captain gave up the chase, and ordered the boats to return. Thus ended our first right whale chase.

December 25th.—This was a day of general starvation and discontent. I had never spent such a Christmas before, and I devoutly trust I never shall again. At sunrise I went to the mast-head. The weather was raw and boisterous, and the sea very rough. I had three hours aloft, after which I was relieved by one of the Portuguese, and went down to enjoy the luxuries of a cold pot of coffee and some hard biscuit. At dinner-time there was no meat for us fit to eat, and the cook had spoiled the "duff." Some of the crew went aft to the captain, and complained that, as it was Christmas-day, we ought to have something to eat; but the captain did not seem to consider any such luxury as eating and drinking at all due to the crew of a whaler; so we were compelled to take a reef in our belts and wait patiently till sup-

Part IV—Excerpts from Etchings of a Whaling Cruise

per-time. We fared little better then, being short of meat, and having tea unfit for use.

****

Things were now in the worst possible condition. Three men had deserted; others had threatened to desert. The captain was terribly out of humor. The mate chuckled in his sleeve, and would have rejoiced had all hands followed the example of the deserters. All this trouble was nuts to him. To weigh anchor for another cruise without our full complement of men was out of the question. We had all sufficiently tested the hardship of whaling with two boats. It was not possible the deserters would be retaken and there were no men to be had ashore except the natives. Still it seemed hardly fair that, with the prospect of a week or two in port, and little to do, our liberty should be stopped for an offense of which others were guilty.

May 25th.—I went below, as usual, after supper. The Portuguese were in earnest conversation. M'F—— and Charley were also talking over some deeply interesting subject. There was a sudden cessation of the conversation the moment I entered the forecastle. It was evident something profoundly mysterious was going on. I inquired what was afloat, but received only an evasive answer, which tended to increase my suspicion. Shortly after I had turned in, Charley came to my bunk, and whispered,

"We have all agreed to refuse duty. What do you think of it? Will you go on deck in the morning, or stay below?"

"Why, what's the matter?" said I.

"Matter enough. We don't want to be slaves any longer. We are determined to have liberty ashore, or weigh anchor and put to sea at once."

The Portuguese overheard us, and joined with Charley, protesting with oaths that they'd go to sea or have liberty ashore, one or the other, and that I'd better join them, if I didn't want to get myself into trouble. I remarked that the trouble would most probably be on their side, and warned them of the consequences which might ensue from a revolt of this kind. They were all in a high state of excitement, however, and would not listed to argument or reason. For my part, I said I would go on deck when called. I was as anxious as any of them to have liberty ashore, but had no particular desire to be put in the fort.

"Then," cried several voices, "you are *a coward!* If you had any spirit, you'd join us; but you're afraid of the captain.

I observed, in reply, that none of them could justly accuse me of cow-

Part IV—Excerpts from *Etchings of a Whaling Cruise*

ardice. I had never flinched from real danger; and I considered it no proof of courage to commit an act of folly, which would only bring additional trouble upon my own head.

"Then you'll sleep on deck to-night! We won't have you here, by G—d!" cried Manuel, the bully of the Portuguese. Juan, Josè, Frank, and some of the others, joined in threatening to put me on deck. I made no answer, but lay still, expected an assault. Charley and M'F——, I believed, would not countenance such an outrage; yet I knew that, when excited, the Portuguese would stop at nothing, however brutal or cowardly, to gratify their animosity; and, even if the two Americans joined me, we could make but a feebly resistance against eight overgrown ruffians, all armed with knives.

After the Portuguese had chattered a while in their own language, they again addressed me:

"You had better go on deck. If you don't, look out to-night!"

To this I replied, that I knew them too well to be intimidated by their threats.

"Then, if you sleep in the forecastle, you sha'n't go on deck in the morning. You can have your choice: go on deck now, or stay below in the morning."

My answer was: "I shall do neither. I have as much right to sleep in the forecastle as you have. Your refusing to do duty is a matter that concerns yourselves. I shall not be driven into trouble by any of you."

These cowardly dogs, who could be so bold on an occasion like this, when they only had but one to contend against, one whom they had every reason to suppose would not fight—for I had never taken any notice of their insults—now began to make demonstrations of an assault. All my past hatred for them seemed be centered in a single moment. I felt as if suddenly inspired with supernatural strength. My blood boiled with indignation and contempt. To use a western phrase, I was, for the first time in my life, really *wolfish*. In the bitterness of intense and loathing hatred I cursed them, taunted them, dared them to lay a hand on me. Now, let it not be supposed that I intend this for bravado; *I knew my men*. I knew that nothing but a bold front could save me the disgrace of being severely thrashed. Besides, I despised them with the most unfeigned cordiality, and it relieved me to let off a little of my exuberant valor. The effect was magical. Not one of them touched me! Charley and M'F—— lay in their bunks chuckling over the fun; nor did their mirth surprise me, for the whole scene had something supremely ludicrous in it. As I turned over to go to sleep,

Part IV—Excerpts from Etchings of a Whaling Cruise

after this bloodless *fracas*, I overheard Manuel say "he had a sharp knife, and I'd feel it before I knew where I was."

About midnight I was awakened by low voices in conversation. To tell the truth, I had not slept very soundly. There was something, in spite of all the bravado of the Portuguese, by no means pleasant in my situation. By listening attentively, I found that Hankley and Antoine, the two Portuguese boat-steerers, were below discussing the topics of discontent. A smattering of their language enabled me to make out the substance of their conversation. It appeared that they apprehended Enos and George would be flogged for attempting to desert. They were resolved the punishment should not take place; and the doubtful point seemed to be, whether, if all the Portuguese united in a revolt to prevent the punishment, Charley, M'F——, and myself would join them. I heard my name mentioned very often, accompanied by epithets of no flattering character; and, indeed, felt rather uneasy till the boat-steerers went on deck.

At daylight next morning the watch came to the scuttle as usual, and called all hands. I had not undressed, so that before any attempt could be made to keep me below, I was on deck. In the course of ten or fifteen minutes the mate came forward, and asked the reason of the delay. Finding how matters stood, he went aft, with an ill-concealed chuckle, to convey the information to the captain.

No notice was taken of the revolt till after breakfast, when all hands were called aft. The captain appeared to take matters pretty coolly, considering the indignity offered to his authority. Had this affair taken place at sea, he would have pursued a different course. There was a consul ashore, however, and he was evidently anxious to avoid having any investigation of the ship's economy before that officer.

"Now," said the captain, very deliberately, thrusting his hands in his pockets, and taking his stand by the main-mast, "you that belong that 'ere crowd, stay where you are; and you that don't, step over to leeward."

No one stepped over to leeward but myself.

"Well," continued the captain, fixing his cold, dead eyes on the Portuguese, "what's the matter? Why don't you go to work?"

There was no answer, till Charley stepped forward and said, "We want liberty ashore, sir."

"*You sha'n't have it!*" was the reply.

"Then, sir, we want to put to sea, and get over the voyage as soon as we can. We didn't ship to lay up in port without liberty."

Part IV—Excerpts from *Etchings of a Whaling Cruise*

"I'll put to sea whenever *I* think proper," said the captain. "Will that suit you? What have you to say, M'F——?"

"I want to go ashore, sir, or go to sea."

"You sha'n't go ashore, nor to sea till it suits my convenience. What next?"

"We'll not work then, that's all."

"Go forward, all of you. I'll soon find a way to make you work. Cook, see that those men have nothing to eat till I give you orders to the contrary! I'll starve this obstinacy out of you."

The mutineers then went forward, and took up their quarters for the day in the forecastle. As it happened, the bread-kid had been replenished that morning, so that they suffered no inconvenience from starvation that day. It was perfectly apparent that, unless they resorted to forcible measures to procure a supply of provisions, they would soon be compelled to yield. Finding the captain resolute in his determination to give them nothing to eat till they went to work, the Portuguese next day came into measure. Charley and M'F——, who were left in a small minority, apparently yielded; but it was with the mutual understanding that they would desert on the first favorable opportunity, and, if no such opportunity offered, they would swim ashore the night previous to the sailing of the vessel. Thus ended the revolt. To my great surprise, I learned, when it was all over, that the captain suspected me of being the originator of all the mischief, and regarded the course I had taken as a mere *ruse* to avoid punishment!

My situation was now more unpleasant than ever. On the one hand, reproached and taunted by the crew for refusing to join them in their revolt, and, on the other, suspected by the captain as the real instigator of all mischief, I had no peace either on deck or below. The prospect of being compelled to spend a year or fifteen months longer on board the barque, with all the horrors of the past increased tenfold, drove me to the verge of despair. I thought of the parting words of the man who had made his escape from the steam-boat at Newport, Rhode Island; I reflected with many bitter thoughts upon my indiscretion in not following his advice; I called to mind the unhappy fate of my friend, left sick and destitute in a foreign land; and, so help me God! sooner than drag out another such year of misery, I would gladly have exchanged my place with that of the most abject slave in Mississippi; nay, so desperate did my prospect seem, that, had the offer been made me to serve a year in one of the state penitentiaries, I believe, from the bottom of my heart, I would have accepted it in preference to the life I now led.

Part IV—Excerpts from Etchings of a Whaling Cruise

In a state of mind bordering on madness, I resolved to take a bold step, which, if successful, might procure me my release, but, if not, might be the means of my disgrace during the remainder of the voyage. I had no confidence in the captain's humanity. I knew very well, if I attempted to desert and did not succeed, I should be seized up and flogged like a dumb brute. The degradation of such a mode of punishment I had not yet experienced, and I was anxious to avoid it, if possible; for it was my firm determination, from the moment I first witnessed it, to take a more summary method of avenging the wrong, should it ever be inflicted upon me, than that of the law. My intention was first to make an appeal to the United States consul, lay the facts before him, and, if unsuccessful, to desert and suffer the consequences. I knew the fact of my appearing anxious to leave the vessel would be a sufficient cause, in the eyes of the captain, to treat me with increased barbarity in case I remained on board; so that if I failed, I might as well desert, and run the risk of being taken and flogged. There were but two Americans left, M'F—— and Clifford. I had no reason to rejoice in the prospect of having, in addition to eight Portuguese, three or four beastly negroes to share the forecastle with us.

I remained on deck till a late hour, reflecting upon my condition. About midnight, finding all the watch asleep, I stole softly down into the forecastle, and, by the flickering rays of the lamp, commenced my work. Seated on a soap-keg, I made use of one of the chests as a desk, and wrote a long letter to the American consul, setting forth the particulars of my unpleasant situation, and the hardships which I had endured, together with the dreadful alternative before me, of remaining on board the vessel another year, if he should not exercise his influence to procure my discharge. I appealed to his humanity—his sense of justice as an officer of the American government. I called upon him, with all the eloquence I was master of, to save me from the horrors of such a life. It occurred to me that the letter might fall into the hands of the captain, or some of his officers, and, to guard against any unpleasant consequences that might arise from such miscarriage, I made no complaint against them; though I could have said a good deal that would have placed them in enviable light. This letter I sealed, and, putting it under my pillow, turned in to dream of home and happier times.

Next morning, perceiving that M'F—— was about to go ashore in the mate's boat for wood, I slipped the letter into his pocket, and begged him, if an opportunity should occur, to hand it with due secrecy to the

Part IV—Excerpts from *Etchings of a Whaling Cruise*

consul. Mack was the very man whom I would have chosen, had I any choice in the matter, for this delicate commission. Honest and kind-hearted, he had my most implicit confidence. I felt quite sure he would spare no exertion to do me a service.

How can I depict the tortures of suspense that I suffered that day? My mind was racked with alternate hopes and doubts. Would the consul receive my letter? What would be its effect? Would he demand my instant discharge, or pay no attention whatever to my appeal? These were but a few of the conflicting questions upon which my mind dwelt during the absence of my shipmate. Hour after hour I watched the boat with eager eyes and a throbbing heart. At length I saw the crew get in and push off. I felt faint with expectation. After a tedious pull they hauled up alongside. Mack's face beamed with good nature. I read in its expression the success of my mission. He had delivered the letter. The consul had read it, but made no comment upon its contents, except that he would be glad to see me the first time I went ashore.

The question then was, would I have an opportunity of going ashore before the vessel weighed anchor? All that day and part of the next I remained in the same miserable state of suspense. At length the mate sang out, "Man the waist-boat!" Every one was anxious to get ashore, and all hands rushed for the davit-tackles. I did not even wait till the boat was lowered, but, with one spring over the bulwarks, made good my place. It fortunately happened that the key of the wood-yard was up at the consul's. While the mate was thinking how he would get it, I volunteered my services to run for it; and the answer had scarcely escaped his lips, when I was off.

The consul chanced to be standing at the door. I approached him with awe and trembling; for he it was who was to decide my fate. Having first delivered my message, and procured the key of the wood-yard, I hesitated whether to address him, or leave the matter entirely to himself. While considering what course I should pursue, he called me, and said,

"Are you the young man who sent me this letter?"

"Yes, sir," said I.

"Well, when will you be ashore again?"

"I can't say, sir; the captain allows us no more liberty."

"Call upon me to-morrow, and I'll talk this matter over with you; I shall see the captain about giving you permission."

Elated with the hope of a speedy release, I hurried down to the wood-yard, and went to work with a zeal that I had not felt for many months.

Part IV—Excerpts from Etchings of a Whaling Cruise

Permission was given me to go ashore next day with one of the boat-steerers, who, as good fortune would have it, was no other than my worthy friend Tabor. I communicated to him the object of my mission. He said he was sorry to think of parting with me; but, for my own good, wished me success. I lost no time in calling upon the consul after leaving the kind-hearted boat-steerer. On my arrival at the consular residence, I was shown up stairs, where Mr. W—— attended to private matters.

"I see," said he, "you are not pleased with the whaling business. You find it, no doubt, different from what you expected. It is a severe lesson to you, and I hope you will profit by it. I am willing to aid you so far as I have the power, but have no authority to demand your discharge from the vessel. If you can procure a hand to take your place, it is probably the captain will consent to an exchange. In that way you may succeed, but I know of no other."

"That, I fear, can not be done," said I; "the captain would keep me, if for no other reason than because I wish to procure my discharge; besides, I owe him thirty dollars.

"I am sorry for it; the whole matter rests with him."

"Then, sir, I shall be compelled to desert. I can not remain on board the vessel another year."

"That would be unwise. We have no other alternative here than to put deserters in the fort, and there you would soon take the fever."

"Well, death is preferable to my present condition."

"If you think it would be of any avail," replied the consul, "you may tell the captain that I will pay your bill; and, to show you that I am willing to do all I can for you, I give you liberty to offer ten, fifteen, or twenty dollars to any of Captain F——'s crew who will take your place."

This was certainly as much as I could have expected from any American. Thanking the consul for his kindness, I set out for Captain F——'s hut, and there made the proposition to those of the Bogota's crew who were not prostrated with fever. It was like offering fire to a burned man. Not one of them would listen to my proposition. They had experienced too many of the hardships of a whaling voyage within a month or two past to be tempted by any pecuniary inducements.

Sadly disappointed, I went off in search of a substitute among the natives. The sun was intensely hot, and the streets were like long, dirty ovens. After a harassing ramble of two hours from hut to hut—seldom meeting any one who could speak English—I fell in with a couple of young

Part IV—Excerpts from *Etchings of a Whaling Cruise*

blacks, who had been to sea before, and could speak English with tolerable fluency. They both seemed anxious to ship, and agreed to take my place if I could make an arrangement of that kind with the captain.

I returned to the consul's, and gave him the result of my search. The only difficultly, he said, in taking natives of the island, would be the consequences it might lead to in case the vessel should fall in with a man-of-war. So many blacks on board would excite the suspicion that she was a slaver, and much embarrassment might result from it. If the captain was willing to take one of them, however, he thought the matter could be arranged.

Without any delay, I proceeded down to the beach, and, finding the boat ready to start, went on board. It was nearly sundown. The work of the day was over, and I found the crew lounging about the forecastle, smoking and chatting as usual. Learning from Clifford that the captain had given orders to the mate to weigh anchor early in the morning, for the purpose of dropping down opposite the imaum's palace to procure a supply of water from the Motoney, preparatory to making immediate sail on another cruise, I became alarmed about my prospect of effecting the exchange. Not an hour was to be lost. As yet I had not broached the subject to the captain.

Plucking up all my courage, I walked aft to the larboard boat, in which he was sitting, and, taking off my hat, respectfully asked his attention for a few moments.

"Well, what do you want?" said he, staring at me as if he had never seen me before. "What's the matter now?"

"I wish, sir," said I, striving hard to suppress my agitation, "to procure somebody in my place. This business does not suit me; I should like to have my discharge."

"What! Discharge, hey? Why, I thought you were very well satisfied. I haven't heard you say you wanted to leave the vessel till now!" replied the captain, with unfeigned astonishment.

"Because I thought it would be useless. Now, sir, if I procure a hand in my place, will you let me go? I have no peace with those Portuguese, and would rather take my chance of dying of the fever than remain on board."

"Oh ho! that's it, hey? If that's all, you may rest easy. I intend putting you, and Clifford, and M'F—— in the aft-hold as soon as we get to sea."

This was a "stumper." I could only reply that, under any circumstances, I would prefer having my discharge, adding, that the consul had promised to pay my bill, and I would engage to furnish a first-rate hand in my place. I was then dismissed, with the remark that "he'd see about it."

Part IV—Excerpts from Etchings of a Whaling Cruise

At five o'clock next morning we weighed anchor and dropped down to the Motoney. Immediately after breakfast, while the men were battening the casks for a raft, I renewed my application. The captain was not at all pleased at the dissatisfaction manifested lately by his crew. He thought he had treated us well, and wanted to know the reason we were all so anxious to leave the ship. It would tell badly for him with the owners, if he went home with but two or three of his original crew. This was the first time he ever condescended to argue the point, and I could not but feel surprised at his wonderful benevolence. However, a little reflection enabled me to see through it. He thought it highly probable I would desert, and preferred having a man in my place, while, at the same time, he would leave a good impression. After some quibbling about having so many negroes on board, he agreed to take one of the natives to whom I had alluded. No opportunity to go ashore occurred till in the afternoon, when the consul's boat came alongside with a clerk, who had business with the captain. I was permitted to accompany him back to the town. In about an hour the captain came ashore for his papers, having concluded to put to sea before officers and all left the vessel. When I produced my substitute, who was a fine, stout young negro, the captain, without assigning any reason, peremptorily refused to take him, and ordered me on board again. I had no alternative but to obey. The barque was all ready for sea next morning, and I felt sure, if I went on board again, I should lose the last opportunity of making my escape, having no doubt strict watches would be set that night. On my way down to the boat, I met one of Captain F——'s men, to whom I had proposed an exchange on a previous occasion. I again broached the subject, and, knowing him to be a reckless fellow, to whom whaling was as agreeable as any other pursuit, I had no qualms of conscience in offering him every inducement. He was a rough, stout-built quadroon from South Carolina, and had been drifting about the world for eight or ten years past. With an iron constitution and a happy disposition, he never troubled himself with any of the niceties of feeling or thought, and I sincerely believed he was much better adapted for the situation than I was. I offered him ten dollars in cash, and all my clothes, together with my sea-chest, and whatever else I had on board, if he would take my place. Such an offer was hard to resist. He said he would not agree to take my place but would ship as boat-steerer, in which capacity he had served on board the Bogota, at the fiftieth *lay*, provided I would give him the then dollars and the clothes. It made no difference to me, of course, in what capacity he shipped, if I could procure

Part IV—Excerpts from *Etchings of a Whaling Cruise*

my discharge by an exchange. I had but little time to talk, however, as the boat was waiting; and, telling him I would consult the captain, and that he must be down about the beach before dark, pushed off once more for the hated barque.

It was decided that we were to weigh anchor and put to sea at daylight next morning. I had but two or three hours to spare; the captain might not come on board till dark, and then it would be too late to make the arrangement which I so ardently desired. I did not know, even, whether he would take my substitute, having already, without cause, refused the man whom he had promised to take.

In the most racking state of suspense I awaited the arrival of his boat. I must have looked like a madman, for the crew made comments upon my unnatural wildness. I was, indeed, somewhat bereft of my senses, and had difficulty in controlling my feelings. A long, weary hour of torture elapsed before the captain made his appearance. With mingled feelings of hope and despair I approached him, and told him the particulars of my interview with the quadroon. It fortunately happened that this was a man to whom the captain had himself applied some days previously, and whose services he was most anxious to secure. The only difficulty was about the *lay*. However, sooner than not have him, he consented to go ashore with me and talk to the man, and if they could agree on the terms, he would give me my discharge.

I need hardly say I sprang into the boat half frantic with delight. So overpowered, indeed, did I feel with joy at the prospect of my release, uncertain as it was, that I could scarcely refrain from telling the captain I considered him a pretty clever fellow, after all.

To make a long story short, the matter was arranged to my entire satisfaction, and we went up to the consul's to sign the papers and settle accounts. I did not feel sure of my release till I held the written discharge in my own hand; then, for the first time in my life, I felt what it really was to be FREE! I am sure I must have made myself very ridiculous. I hardly know what I said or did, for I was delirious with joy. In the phrensy of my delight I shook hands with the captain, and wished him a pleasant and prosperous voyage; well knowing, even then, that for half a dollar he would have sold my liberty for life had it been in his power.

Once more I returned to the vessel to bid good-by to my old comrades. The cooper, Tabor, the mate, Charley, and M'F—— shook my cordially by the hand, and wished me all success and happiness. I really felt sorry at

Part IV—Excerpts from **Etchings of a Whaling Cruise**

parting from friends to whom I had become so warmly attached, more particularly Tabor and the poor Yankee boy, for both of whom I entertained the strongest regard. Poor Mack! I had been his warmest friend, and the tears stood in his eyes as he bid me good-by. I would have given the savings of ten years to come to have had him share my good fortune. Nor was it without emotion that I parted with the Portuguese. We had lived many months together; we had endured the same hardships, faced the same dangers, suffered together, and stood night watches together; and, although I did not regret that the tie of our fellowship was broken, I sincerely wished them well, and was deeply impressed with the thought that, as our paths through life lay in different directions, those faces so familiar to me, I should, in all likelihood, never see again. From the bottom of my heart I forgave them all I ever had against them. They had enjoyed but few of the blessings of social life; their career from childhood had been one of labor and hardship. I saw more in them to pity than to blame, and I felt sorry for every harsh word I had spoken to them during our unhappy voyage.

The transition from the filthy forecastle of a whaler to a large and commodious stone house, furnished in European style, and to the society of a little circle of educated New Englanders, was so unexpected and so difficult to realize, that for many days I could hardly avoid thinking it a delightful dream. How my blood thrilled through my veins and how my heart bounded with joy, when I saw around me objects that brought to mind home, friends, civilization, and all the blessings and comforts of social life! Above all, how entrancing, how indescribably delightful, was my long-wished-for FREEDOM! how happy the thought that I was no longer subject to the whims and caprices of a tyrant!

So great was my horror of the life I had led during the past year, that in my sleep, for weeks after my release, I struggled with imaginary foes, whom my morbid fancy pictured in the act of dragging me on board again; and often, when I awoke and looked around me, I wondered what had become of the rusty forecastle lamp with its dim, flickering rays.

Part V: Excerpts from *The Gam* by Capt. Charles H. Robbins

When two whaling ships happened upon each other at sea far distant from their home ports the meeting was called a gam in seafarer's language. It was a special occasion on isolated voyages, marked by an exchange of news from home and the latest whale sightings, and trading food, reading materials and letters for transit to the nearest port.

The ships' officers and crew briefly came together, often playing music for each other, and if a whaleman had the gift of storytelling would captivate his fellows with yarns and other tales. Capt. Charles H. Robbins had spent most of his life in the whaling trade, first going to sea at the age of fifteen. Toward the end of his life at the end of the 19th century, he collected a number of stories into a gam with his readers, opening his life and his world to us. After his death at 81 years of age he was called a "type of New Bedford whaleman. That he was a courageous, valorous man, a man of hardy daring we need not say, but withal, he was a gentle, kindly, conscientious God-fearing man, excelling in character." A few of his stories first published in 1899 are included here.

Making a Master

One keen winter morning in 1837, there stood before Captain Lewis Tobey, of the ship *Swift*, a boy asking permission to go with him on his next voyage. The captain looking sharply at his visitor, saw a lad of hardly fifteen, slight, erect, with dark hair and deep blue eyes, and something about his square chin and firm mouth which he seemed to think argued well for the future.

"Take your hands out of your pockets, and tell me your name," demanded the master.

"My name is Charles H. Robbins," replied the boy, his eyes on the hands which were now clasped in front of him. That boy was myself.

"Who is your father?" was the next question.

"My father was Lemuel Robbins, sir," I replied softly. "He died six years ago."

The captain remembered that he had seen the boy before. Many evenings when he sat telling sea tales to the friends who were his entertainers on shore, this lad, visiting his chum, the young son of the family, had been among his listeners.

The result of that morning interview was a decision that I should go as cabin-boy on the ship *Swift*, which was very soon to sail.

As I walked away, after my conversation with the captain, any unexpressed feeling was that for me real life was about to begin; the life of achievement, of accomplishment, of profit. I had made what I considered a good bargain with the ship-master. I was to receive fifty dollars before sailing, and fifty dollars on my return home. What boy does not regard a hundred dollars as a fortune?

There was pressing need that real life should begin, that good bargains should be made. Nine living children had my mother, the widow Robbins, of whom I was the seventh. My birthplace had been Mattapoisett, where I had lived six of my fifteen years. The little New Bedford house on hilly Foster Street, which had been my home for nine years, was uncomfortably full, and insufficiently furnished with life's necessities, and there were only two men children among my mother's brood. To the *Mercury* office, where I had been employed as office boy and paper-carrier, there often came a marine reporter, telling, with the glib tongue and ready imagination common to his kind, of the adventures and the gains of those who secure a livelihood by the sea. These relatings, to which were added those of Captain Tobey, aided by my own imagination, convinced me that no better employment could befall a man than was to be had on ship-board. To the conviction that it was my duty to materially aid in supporting our family, was added youth's longing for change, and the appeal which the wide, free ocean life made to my poetic temperament. To my mother's opposition to the contemplated voyage, I said, "If I don't go now, I shall at some future time," and her consent was finally given.

One February day I went out of the old home gate a very sad feeling boy. The very squeak of the hinges seemed to be saying good-bye. I was going to join the ship. I was already homesick, but even with that terrible feeling tugging at my heart I did not wish things otherwise. The remembrance of the kind tones and genial manner of the captain I had so many times met on shore was a great comfort to me in my loneliness, as with misty eyes I looked farewell to every well-known object as I hurried down the familiar streets which led to the wharf. But alas for my hopes of sympathy! As having reached the ship, I stood ready for my duties at the foot of the stairs leading below, the captain came on board carrying in his hand a bandanna handkerchief which, with its ends tied together, served as a receptacle for half a dozen fine apples.

"Here!" he shouted to me, "take these and carry them to my cabin, and if you put one of your teeth in them I will break every bone in your body."

Thus early and abruptly I was made aware of the difference between an old school captain ashore and afloat, and reminded of the remark of a previous cabin-boy of the *Swift*, who had said to me, "You would enjoy hell better than a voyage with Tobey as master."

For ten days the *Swift* lay in port, hemmed in by ice, and I, so near home and so unutterably homesick, was not allowed to leave the ship. The confined sailors, the liquor which they had brought on board as a part of their outfit becoming exhausted, determined to cut away a boat and land at all hazards, but on its becoming known that the mate had given orders

Part V—Excerpts from *The Gam*

to shoot any sailor who attempted to do this, the men gave up the idea. At last, fearing that longer delay would mean mutiny, the captain had a passage cut through the ice, and the ship cleared the harbor.

Then began that drill which transforms raw sailors into experienced seamen. The first thing which a newly shipped hand is required to do is to learn the rigging. Every mast and spar and sail and rope must be so familiar to him that in the darkest night, without a ray of light, he can handle them quickly, accurately and effectively, knowing everything as well by touch as by sight. To many a newly made sailor it is one of life's most terrible moments when he is first ordered to the masthead, that cruelly-tapering, suggestively-towering spire, a glance at whose skyward reaches causes the brain to reel and the heart to fail. But the mounting is inevitable. If one refuses it, punishment is also inevitable, and the task still to be accomplished. To a kind hearted master this forcing of young sailors into the rigging is an almost heartbreaking experience; to the calloused captain a most exasperating one. But the *Swift's* cabin-boy wanted to go into the rigging, longed to learn everything, great and small, about a vessel, and that right quickly. As the days went on, although my duties as cabin-boy were never neglected, I worked much of each day among the men, constantly gaining a thorough knowledge of every form of seamen s craft. In the boats which were sent out to drill for the capturing of whales, I became an expert. While the captain was trying to teach the second and third mate navigation, I was sitting in a state-room opposite the cabin occupied by the three, noting all the many times repeated instructions, and at the end of the lesson I went below and worked out the correct solution to the problem, which I handed to the surprised and delighted master, who thereafter taught me navigation with great thoroughness, and depended upon me to keep the ship's time.

It was continually being made evident that this captain of ready blows and fierce language had yet a place in his heart which could be touched by real merit and conscientious devotion to duty, and I often found that presents followed kickings, and the imparting of valuable knowledge or the granting of privileges a torrent of abuse.

So rapid was my advancement that during this first voyage, as I became boat-steerer, I was mentally, morally, and physically becoming the well-rounded man. Many a cruise as mate and master did I make after this initial one, but that fifty-four months of untold hardships, of uncounted humiliations, gave me the readiness, the knowledge, the experience, and the resolves from the fusion of which there was evolved the captain of my dreams.

That Great Leviathan

Pleasantly my memory runs back sixty years to the day of our departure [on the whaling ship *Swift*]. Stars and stripes at the peak, Blue Peter at the fore; officers and crew on board; four boats on the cranes; and the hold filled with white oak casks and a stock of provisions to last three years and more; then, as somebody or other says, "Waiting is what?"

Waiting is the pilot. But, once aboard, his majesty takes command, and the voyage is begun. The captain, while the pilot remains in the ship, is a mere inactive looker-on, a person of no more consequence than a passenger or a spare figure head.

"Mr. Mate, are you all ready?"

"All ready, sir?"

"Then heave ahead!"

"Aye, aye, sir. Man the windlass!"

Now comes the confusion, the hurrying and blundering, invariably seen on board a ship when she is getting under way with a crew made up largely of green hands; then an attempt on the part of the officers to bring something like order out of this hurly-burly; and while at every turn of the powerful windlass the chain cable rattles heavily on deck, the *Swift* walks steadily up to her anchor as if impatient for the word to spread her white wings and be away.

"A-vast heaving," shouts the first officer. "A short stay peak, sir!"

"Aye, aye," responds the pilot.

"Let fall and sheet home top-s'ls and to'-gal'n-s'ls!"

Away sprang half a dozen men aloft, and soon the broad sheets of canvas are unfurled and hauled home and the yards are mast-headed.

Next a volley of incomprehensible orders :

"Brace head yards a-starboard!"

"Lay helm aport!"

Part V—Excerpts from *The Gam*

"Heave up the anchor!"

The first mate answers, "All away, sir!" and you know then that the good ship has loosed her hold on terra firma, and you watch her movements, as gracefully as a girl in a minuet she turns her head seaward.

The pilot springs to the bow, now and again shouting his orders to the helmsmen, who invariably echoes the words, that there may be no possibility of mistake.

And so, with a breeze fresh and free, we sped down the bay, borrowing a little, now on one shore, then on the other, or shaving close to some rocky ledge, as our sharp-eyed, skillful guide might direct, in order to shorten our course from the confines of harbor to the freedom of the open sea.

A little farther, and we open up Gay Head lighthouse on the western end of Martha's Vineyard, so called from the abundance of wild grape vines growing there. Once outside, the tiny pilot-boat, which has been dodging about the heavy ship like a will-o-the-wisp, shoots alongside and his lordship the pilot and our friends, mostly men of the sea, hasten to make their adieus, and descend to the restless little craft that will soon take them back to their homes. The lingering grasp of hands, the ill-concealed tremor of fare wells, and the moistened, glistening eye, tell of the friendship of men who have together battled with the giant seas and fierce winds of the Horn, who have stood shoulder to shoulder when shortening the wings of their hurrying ship in the short-lived gales of the Equator, and who have for long years shared alike in common hardships, joys and sorrows.

The little fairy shoots ahead, and, flying up into the wind, is soon on our weather beam, homeward bound. Three rousing cheers from her deck, and three from the outward bound, and we are alone on the sea, with nothing binding us to the shore but memories of the past and hopes for the future!

And now, indeed, though with everything yet to learn, I was fairly made a sailor of. There was no possible back-wending, however I might thereafterward mope and whimper. Accordingly I turned my heart manfully toward my strange, new life and faced it with earnest cheer.

The first day out, the ship's crew is divided into two watches, larboard and starboard, the former always headed by the first officer and the latter by the second. The men are mustered aft and the rules of the ship laid down to them. At seven that evening, the watch is set, the second officer always taking the first night watch from the home port, and those not on duty go below and sleep if they can. Next morning all hands are called aft again, this time for choosing boat's crews. The first officer takes precedence by selecting one man, followed in turn by the second, third, and fourth mates, each choosing one, until every boat has a crew standing by her side. Then follows, usually, the emphatic caution, "Now remember to which you belong, and bear a hand when she's called away!"

And what of the voyage? Southward? Yes, in the main; crossing the Gulf Stream; battling, stripped for the fight, with many a heavy gale; passing, with men all the while at our mastheads, through the "horse latitudes;" lowering our boats, now and then, to give our whalemen practice in rowing; and taking advantage, now, of every slant of wind to press on our way toward the stormy Horn.

Days and long weeks go by, nor are we alone in the tedious struggle. Several sails are in sight, all striving to get south.

And so, with bracing round, or squaring the yards, making and shortening sail, and backing and filling generally, we get a sharp squall, with rain, from the eastward, and then the old salts cast at each other significant glances, which, if rightly interpreted, would say, "I believe we've got the Trades at last!" After a few hours, the wind moderates and hauls to the northward. All sail is set again, with the breeze fresh and free, and we go bowling along to the southward at the rate of ten knots an hour.

Oh, the beautiful world of waters! Almost every day we pass ships showing the flags of different nations, some near and others in the far distance, all under a press of canvas, and all seeming to revel in the bright sunshine and the breeze. The water, too—so warm and so transparent—is full of life. Porpoises, dolphins, albicore, and barricota are gambolling and sporting in the summer sea. Thousands of birds are on the wing or resting on the waves, while not infrequently a huge fin-back, or sulphur whale rolls lazily along, now throwing clouds of misty spray into the air, and again lashing the water into foam with its broad flukes, doubtless to rid himself of the numerous parasites which persistently strive to fasten themselves upon these worthless vagabonds. Vitality and loveliness are above, beneath and around us, and we seem verily to be sailing on a sea of enchantment. The stars seem nearer, and shine and twinkle with that wonderful brightness seen only in that southern hemisphere. The North star has dipped into the ocean, not to rise again until we cross the Equator on the Pacific ocean. Instead we gaze in novel delight upon the Southern Cross, and we are constantly looking for that mysterious and ghostlike thing known to seamen as the Magellan cloud, and said to mark the entrance to the famous straits of that name. It is enough to make a man quote the spirited lines of Kipling:

> "O, the blazing tropic night when the wake's a welt of light
>   That holds the hot sky tame,
> And the steady forefoot snores through the planet-powdered floors,
>   Where the scared whale flukes in flame!"

Round the horn we fly, wrestling with giant seas, and then, while penguins and fur-seals go sporting and harking around the *Swift*, we pass the rugged, half-glaciered island of Tierra del Fuego. Warmer, day by day, grows the air and softer. At last, though never a spouter have we yet raised out of the ocean, our hog-yoke tells us we are upon the rich off-shore whaling grounds.

After we had been out from home eight long months we chanced to speak the full-rig ship *William Rotch*, and I then beheld a sight that stirred my soul from truck to keelson and knocked my youthful emotions galley-end wise. For the *Rotch* had a monstrous whale, just taken, tethered alongside.

There he lay, a bit ingloriously, to be sure, for he was riding belly uppermost and tail foremost; but I felt like a Titan when I looked at him. That was the prey I had gone a-seeking. I was a fighter of dragons and worse. Oh, what more heroic opportunity is offered to man or boy than to join battle with such a monster as that? So thought I (turning sea-green the while with envy of yonder lucky crew) and longed, with inexpressible heart-hunger, for our own first whale-fight. Moreover, I wished myself at that moment a blood-thirsty pirate; for, ethical considerations aside, it would have been a gratifying relief to my feelings had we boarded that ship, like "gentlemen of fortune," bowie-knifed her gallant crew, and stolen that whale away.

We kept company with the *Rotch* all night, and we "gammed"—that is to say, we exchanged visits back and forth, and enjoyed a general fo'c'sle pow-wow for'ard while the officers made merry in the cabins; and particularly merry they were that evening, too, for the old man s brother was mate of the *William Rotch*, and the two had not come face to face for many an eventful year.

But who knoweth what a day may bring forth?

The sun came red and fierce and savage out of the water. The morning mist lifted lazily off the ocean. The long-expected happened.

That Great Leviathan

Try how I will, I cannot recall in any former or any subsequent experience, whether upon land or sea, such a panic and stampede of emotions as instantly followed a ringing cry from the mast-head.

"There she blows!" I heard a man shout.

A haze seemed to rush over my soul. All that happened in the next five minutes is an utter confusion of tumultuous and ungovernable impressions. "All hands" must have been called, but I could not hear the words. Every man sprang toward his boat—in fact, the movements of the crew were automatic and inerrant—yet I made nothing coherent of their desperate hurry. Almost in an instant the boats were lowered swift away; but not until three long whale-boats were dashing out after the great leviathan and bent now upon actual chase, did I come to myself far enough to take good account of how this vast concern was being brought to pass.

I have heard of buck fever. But, lands and seas, it is nothing to whale fever!

Nevertheless, in the midst of so crazed a mood, I did, without so much as considering it, my appointed duty which, for all that, was not difficult; being, as long beforehand I had been instructed, to remain on board and do nothing. That was a simple task, but by no means agreeable.

It was certainly a vivid contradiction, as I have often since reflected, that while I, who was least in the struggle, went clean daft for the moment, the whale, who was of all concerned most gravely implicated, lay spouting contentedly only a small way from the *Swift*, and as wholly free from worry or care as a comfortable cow nibbling pink and white clover-tops.

"Boy," said the cooper, for he stood next to me and together we watched the chase, "I'll bet my go-ashore shirt and pantaloons they'll set you a-turning that 'ar grin-stun!"

This sage observation was the expression of a splendid optimism, for when a whale is being cut in, the cabin-boy turns the grindstone while the cooper sharpens the cutting-spades.

"Oh, by Reuben Ranzo!" yelled the cooper, grabbing me by the collar, "They'll galley him!"

Then, tightening his grip on my neck till I thought he would strangle me, he emphasized his sudden plunge into pessimism with a blast of emphatic and unmistakable English.

Luckily for my continued existence, the fortunes of the whale-chase suddenly grew brighter. The cooper loosed his unconscious grip on my throat and leaned out over the rail, his eyes bulging with intense interest.

Part V—Excerpts from *The Gam*

The chief mate's boat approached the column of steam that rose from the whale's spout-hole.

The harpooner hurled his merciless iron.

The iron took hold in the quivering flesh of the whale, and instantly the captain's boat dashed up and a second harpoon went hurtling through the air to plant itself close to the first. The whale writhed with sudden pain and fright, but did not go down. He preferred to fight.

The old man, however, had plans of his own; he would kill the whale, and that immediately.

He bellowed a hasty order to the mate, thinking to drive the mate's boat out of his way, but he had not calculated upon the stubborn ambition of that hot-headed officer. The mate never budged.

Enraged at his opposition, the captain crowded in between the mate and the monster, and ran his lance into the whale's vitals. Then there was such a commotion as I had never before witnessed. The whale went into a frantic flurry, barrels-full of rich, dark blood were hurled into the air from his spout-hole, the boats dashed away from him as they would from an enraged sea-serpent, and behold—a half-dozen men floundering about in the water!

"Stoven!" yelled the cooper, renewing his unconscious assaults upon my collar. "Served him dead right, I swear! An' bless ye, boy, the old lobster-back can't swim a stroke!"

Indeed he could not. There was the captain in the water, as helpless as a lady, and two of his men were trying their best to keep him from sinking, while one of the two uninjured boats was coming up to take him aboard.

"Same old yarn," said the cooper. "I've sailed with the old man five year if I've sailed a day, an' I tell ye, boy, he's done this lubberly trick forty times over. Gits wearisome, now an' then, dead wearisome for them Jacks to float a poor lubber that won't learn swimmin', and dead wearisome for poor old Chips to have to mend the old man's boat after every blessed chase."

"Then why doesn't the captain learn to swim?"

The cooper ventured no answer. He was watching the mate getting a line fast to Old Blubber. Suddenly he bethought himself of grindstone and spades, and as quickly was off to make ready for the work that would turn me into a slavish minion.

Even before the boats had come in and had got the whale alongside and well into the fluke-chains, the grinding of spades began.

That Great Leviathan

Often and often I had heard men of the sea tell how a whale was cut in and tried out, but now, with my own lucky eyes, I was to see the thing done.

But before I describe how the whale was cut in, I must say something about whales in general.

There are many kinds, but only two are of importance to whalemen. The right whale is sought for his bone. The sperm whale is sought for his blubber. We of the *Swift* were sperm whaling.

Pictures of whales are uniformly deceptive. They give the impression that a good part of the animal (not fish,—a whale is a hot-blooded mammal) can be seen above the surface of the sea. They also indicate that a whale's spout is made of water. It is no such thing. All you can commonly see of a whale from the ship's deck is his spout and that is a mere column of vapor. It's his breath. Get that once in mind and you'll never call a whale a fish. You never saw a fish breathe air. You never found a fish warm enough to belch out white vapor on a summer's day like a steamboat.

Such, then, is the whale's spout. And by the spout the two kinds of whales, sperm and right, can be distinguished. A sperm whale has but one spout-hole, and throws the spout forward at an angle of about forty-five degrees—a thick spout and not very high, rising from a point near the whale's "nose." A right whale has two spout-holes, very close together. They are about eighteen feet from the end of his head and, of course, much nearer his lungs than is the case with the sperm whale. Consequently the vapor shoots up higher and as straight as a mast. It spreads as it rises. I suppose, too, that the bigness of a whale is something few landsmen could well give account of. As a matter of fact, a sixty-foot whale is about as big as you will ever see. Big enough, says any whaleman—to serve as a very worthy adversary to pigmy man who goes to slay him!

Very naturally you ask, as Brutus did (or was it Cassius?): "What meat has this, our Caesar fed on, that he is grown so great?"

That depends on your whale. The sperm whale, having teeth, lives on deep-sea jelly-fish. The right whale, which is as toothless as any dotard, lives on a tiny red creature called brit, no larger than a spider, but so numerous as to color the water a yellowish red over whole acres.

It is because of his choice of diet that the right whale has his mouth filled with a huge sieve of whalebone. That sieve is to let the *brit* through and to shut bigger sea-things out.

The arrangement is a decided success. I have seen a right whale make a scoop of his broad lips and rush through a field of *brit* (like a snow-plow through a drift) and leave a trail of blue water behind him. That is a sight to remember and also a sound to remember, for when a right whale is feeding he spouts with tremendous force. At such a time you will have no hope of striking him.

But right whales don't concern me nor do I concern right whales. We were after oil and we wanted sperm whales or none.

The oil is made from the blubber, mainly, and the blubber covers the whale like a thick coating of fat pork. In one sense it is a blanket; it keeps the whale warm in the coldest sea-water. In another sense it is a shell—or even a padded coat; it relieves the tremendous pressure of the water upon the whale's body when he sounds to the depths of the sea.

Sperm whales have, as already intimated, their ups and downs. A large sperm whale remains under water from forty-five minutes to an hour and a quarter. That is a fact to go by. When a whale has sounded and you are waiting for him to come up, it is a relief to know that some sort of limit is set upon his delay. But that is not all. You can judge where he will come up. For a whale travels, unless vigorously disturbed, about two miles an hour. So you note which way he headed when he sounded, and you measure off two miles in that direction, and you know where to meet your friend again. This is an infallible rule whenever it works.

But a whale has something beside ups and downs and blubber. He has a marvelous sagacity. By some mysterious process, which I suppose the Society for Psychical Research would call "thought transference," whales pass the news of disaster from one end of a school to another. When one of the company is wounded, every whale within a radius of four miles is advised of the fact. Sometimes the alarm will bring speedy assistance. That gives the whaleman only a better chance to ply his gainful trade. Sometimes a retreat is ordered. The whole squadron will dash away as by some instantaneous common impulse, evidently terror-struck.

Can a sperm whale be called a globe-trotter? Be that as it may, the sperm whale migrates far and wide. Ships cruise on the shores of Chili and Peru at a distance of from two to one hundred leagues from the shore, and you will often see both in- and off-shore vessels doing nothing. At other times all will be engaged. Where were the whales while the ships lay idle? Roving over the broad seas, no doubt, and many a mile away, a-taking of their ease.

It is known to a solid certainty that whales have been harpooned in

That Great Leviathan

the Atlantic ocean, and have been afterward taken in the Pacific. The marks on the irons proved the identity of the whale every time. Old Blubber seems to travel for change of scene. It is clear that he is not led to migrate by any fear of the whalemen. Indeed, whales are not easily driven away from their feeding-ground by ships.

But whatever the ups and downs of that whale alongside the *Swift*, and whatever the vicissitudes of his travelling days, one thing was clear. That whale was dead. Like Marley of blessed memory, he was dead as a door-nail. Unlike Marley, however, he could never come to life again. They were cutting him in. I saw it done.

I beheld two stages slung over the side of the ship, each stage six feet long and a foot wide. Men stood upon the stages with sharp spades one to cut the blubber, the other to kill the sharks that would have devoured our prize.

I saw an aperture made near the whale's fin. I saw the great hook inserted. I saw a semi-circle cut around the hook.

Then they took the falls to the windlass. The windlass wound in the falls. The falls passed through a block at the main-mast head. The falls then became the tackle, heaved hard at the iron hook, and stripped the blubber from the whale.

The blubber came off in a continuous spiral strip. The whale meanwhile kept turning over and over in the water. The ripping of the blubber from the carcass was guided by the sharp spade of the officer on the stage.

I saw a strip of white, pork-like blubber, twenty-five feet long and five feet wide, hoisted into a perpendicular position and its top touching the mast-head. Then they cut the piece ("blanket-piece," they said) loose from the whale and lowered the blubber into the ship's hold between decks, at the same time attaching the other tackle to a fresh cut in the whale's flesh and preparing to raise another blanket-piece.

I saw this process repeated until the blubber was stripped from the whale.

I saw the head cut off from the huge beast and hoisted on deck. I felt the great ship strain. The standing-rigging on the starboard side slackened. The mast bent over like a whip-stock. I saw the *Swift* listed till her plank-shear was nearly level with the water.

A filthy column of black smoke rose out of the try-works. They were cutting the blubber into horse-pieces, mincing these pieces, and putting the hashed blubber into huge pots with brick flooring under them and a blazing fire of blubber scraps blazing around them. Thence the oil passed into a huge copper cooler and thence in turn into casks.

They made merry over the boiling. They nibbled bits of fried blubber, and they fried doughnuts in the grease. The whole ship was befouled, but we soon had her cleaned up again, man-o'-war fashion; and what was better yet, we coopered a hundred barrels of oil.

"Lands and seas," said I to myself, "this is the biggest business afloat or ashore."

But as yet I had not chased a whale.

Beach Combers

During our cruise around the Navigator Islands, and while we were sailing along the coast of Upolu, hugging the shore—country-whaling, as they say—a canoe came off with several natives and a white man as interpreter. They were anxious to trade, offering fowl and fruit in exchange for cotton cloth.

While the traders were on board, the man at the masthead sang out, "There she blows!" and sure enough there was a school of sperm whales, cows and calves, going to the leeward. We kept off for them, and, in doing so, passed another canoe, steered by a white man, who had put off from another village to trade. Under the circumstances, the old man thought best not to heave to for them—whales won't wait, and traders will—so they went away for the land, the white man cursing and swearing and threatening to seize and hold our boats if we attempted to land at their village.

We came up with the whales, lowered our boats, and were soon fastened to three. In a short time we had them turned up and dead. We left them with a "waif"—that is, a small flag—stuck in each, and it then became the duty of the officer on board to take the whales alongside, while we went in pursuit of others. It was not long before two boats, the captain's and the mate's, struck again, and soon had the old flukers spouting blood. The first and second mates were left with these two, while the captain went on board ship. There he asked the third mate, who was on board as ship-keeper, in what direction the three whales were which we had waifed. The poor fool could not tell. All he could say was they were to windward, and he had passed them. The old man was enraged at this reply, there were three fine whales, which, if the third mate had done his duty, ought to have been taken alongside and made fast with fluke-chains and towed after the boats in chase. As it was, we found only one of the waifed whales, for it was near

Part V—Excerpts from *The Gam*

dark when we began to search for them. Three whales, however, were already ours, and together they made seventy-five barrels of oil.

After stowing down the oil in the hold, we stood in for the island and went on shore to trade. And now we feared the consequence of passing that canoe without allowing her to come alongside, for we had none of us forgotten the threats and curses that promised us trouble if we landed at their village. The beach-comber tried to have the islanders seize and detain our boats for not having allowed them to come aboard. Nevertheless, we made bold to go ashore, and it was only by sheer luck that we ever got off alive; for we happened to have among us a fellow who understood the native language, and who overheard the islanders hatching up a murderous scheme to "do for" the whole lot of us. This chance interpreter ran and told the old man, who ordered the boats to put off to the ship. Every Jack tar of us knew that that meant stepping lively, and we wasted no time. We launched those boats as quick as coast-guards, but we were not a moment too soon. For no sooner had we bent to our oars than the whole mob of savages made a rush for us, yelling like a pack of lunatics, and running into the water after our boats, but not succeeding in their attempts to get hold of them. The whole brutal job was the work of that villainous beach-comber, and shows how low a white man will get when he sells his birthright and goes to live with savages. The worst insult you can offer an able seaman is to call him a beach-comber.

When we came home after that long, long voyage, we had with us a boat-steerer from this very island,—no beach-comber either,—and thereby hangs a tale. He was the only survivor of two boats crews of twelve men from the full-rigged ship *William Penn*, of Falmouth. The *Penn's* boats were captured and their crews murdered, all but this one man. He was on the island several months eking out existence horrible existence, and expecting his life might be taken at any time. It was on a previous voyage in these waters that our captain found the fellow on the island, and helped him to make his escape. He had shipped again with the old man after that voyage was ended, and was still with us.

This is how the captain took him off. The ship was cruising near Upolu, and the old man sent in boats to trade, going ashore in one of them himself. On the beach he met this sailor, who told his story and begged the old man to help him get away. The thing had to be accomplished by stealth, for the islanders would have butchered the whole crew if they had got wind of the plan. There was a point about a mile from the village which made out some distance into the sea. This man was to go there in the night, and a boat was to be sent from the ship to take him off. It was a life and death venture, for the reef stretched away out into the sea, and the surf came pounding down on those rocks with a roar like thunder. No boat could ever live in such a sea, and there was no hope of landing without smashing your boat to match-wood. There was just one thing to be done. The ship's boat must lie well off the reef, and the man must plunge into those bellowing breakers and swim for his life. If he made it, well and good. If he didn't make it, why death on the reef would be luxury compared with death at the hands of those blood thirsty wild men. So the terrible risk was taken. The daring rush into the foam, the desperate fight with the breakers, the long struggle in the dark, and, at last, life and liberty! They picked him up and took him aboard the ship. He remained on her and came home with her; and he felt so grateful to his deliverer that he shipped with him for another voyage, which was his last. For then he married a Boston lady and lived in that city until he died, not long since, leaving a family to mourn his loss. He was an esteemed friend of mine, and his spirit has gone out, I trust, into the Better Land.

Right Whales

We cruised about three months in the Southern Ocean, looking for right whales. We saw many, and took six hundred barrels of oil and about five thousand pounds of bone.

One day, when the weather was fine and the ocean very calm, we lowered and gave chase to two monstrous right whales that we're going slowly to the leeward. The captain's boat came up to them first and succeeded in striking one. Instantly down went both the whales. When they came up again, the mate struck the other one. They proved to be a bull and a cow—the cow was struck first.

The bull made the sea foam. He cut around in great fury and stove two of our boats—the captain's and the mate's—and the lines had to be cut to get clear. The second mate came along lively and picked up the crews, which came near sinking his boat. Eighteen men in one boat, and the ship four miles away to leeward—a pleasant prospect! And as the wind had died down completely there was nothing for it but to row, and that in an all but sinking boat, so crowded you could hardly move without knocking your neighbor overboard.

But that was not the worst of it. The worst of it fell upon myself and another dare-devil young chap—or rather he and I brought it down upon ourselves, for we volunteered. It was this way. The captain was bound not to lose sight of the stoven boats, and wanted two of the men to stay by them until he could bring the old hooker and pick them up. We two, being young and fearless, offered to take the job. We stood each on the stern and bow of a boat, sunken just to the water's edge, and hung on to a flag-pole for three terrible hours, with the two wounded whales cutting about and making the water white with their huge flukes, only a little way from where we stood.

All that while we were afraid for our lives, as we were out in the middle of the ocean and the ship was four miles off.

It is always with a shudder that I recall that adventure, though fifty years and more have gone by since then. But I remember that even when the danger was worst, we found room for joking, and one of our men cried out, "Better have paid your washwoman!" That is the usual gibe when a man is caught in a stoven boat, for there is a belief among whalers that if you don't pay your washwoman you'll suffer the penalty of getting your boat smashed.

Whale-Land and Its Customs

There are many kinds of whales, and as much aristocracy among them as in European society. The king among them, and, indeed, of all the Sea's inhabitants, is the Greenland, or right whale. This sovereign is sometimes seen on the coast of Britain, and occasionally in even more southern latitudes, but its favorite and most occupied quarters are the northern seas, chiefly in the Arctic regions.

Of its grotesque and unsymmetrical body; which is often from sixty to seventy feet long, the head occupies from a third to a fourth of the entire length, the right side of the skull being larger than the left. It has two small fins, or flippers, but its progress is made by means of the tail, which is five or six feet long and twenty feet wide. The dirty looking, almost entirely black, skin is naked with the exception of a few bristles about the jaws, its surface being moistened by an oily fluid. The lower surface of the true skin extends into a thick layer of blubber or fat, which is from a foot to two feet in thickness This blubber fortifies the animal against cold, and, by rendering the body much lighter, helps to resist the pressure of the water in great depths. The eyes, about the size of those of an ox, are situated on the side of the head and have very acute sight. The spout-holes of the right whale are from eight to twelve inches long and comparatively narrow, and are situated on the most elevated part of the head. The powerful tail can shiver with one blow a large boat to splinters, or toss it and its crew a long distance into the air. The plates of baleen, or whalebone, suspended from the roof of the mouth number three to four hundred on each side. The base of each plate is embedded in the membrane that covers the palate, the edge forming a loose fringe composed of pliant bristles.

The Rorqual is of the same family as the Greenland whale, but a sort of poor relation, as it is despised and rejected by whalers unless it is the only prey they can seize upon. He is larger than the Greenland whale, sometimes measuring a hundred feet, and is a sort of slaty gray and whitish beneath. Like his haughty relatives he is found for the most part in Arctic seas. He feeds on large prey, his throat being much more capacious than that of the right whale.

If the Greenland whale is king of the deep, the sperm whale is prince. This whale is black, and is found in nearly all seas, but most frequently in southern hemispheres. What sailors call the "bull whale," an ancient male, has a large gray spot on the front of the head. The throat is large enough to admit the body of a man. The mouth has no baleen. The upper jaw is without teeth; the lower jaw having twenty-five or thirty on either side, according to the age of the animal, which are conical and slightly recurved, deeply embedded in the gum, from which they reach for about two inches. The lower jaw is extremely narrow, the teeth fitting into cavities in the upper jaw. These teeth weigh from one to three pounds each. This whale has a single spout or blow-hole situated near the front of the head. The enormous head of the sperm whale is mostly occupied by a cavity in front of and above the skull, called by whalers "the case," which sometimes holds as many as fifteen barrels of spermaceti. This substance hardens on cooling. The oil with which it is mixed in the case is separated from it by drawing and squeezing, and the yellow unctuous spermaceti becomes the beautiful, pearly-white, flaky substance of crystaline purity which we know. This oil gives to the fore part of the whale's body a lightness which enables the animal to float and rest.

The "junk," a thick, elastic mass which occupies the head under the case, also yields a considerable quantity of sperm oil. The substance called ambergris, which is used in the most expensive perfumery, and in some Catholic and Mohammedan churches as incense, is found in the intestines of diseased sperm whales, and is supposed to be generated by indigestion. It is extremely rare, and often commands the astounding price of three hundred dollars a pound.

There are many other kinds of whales, but they are for the most part imperfectly classified, and of little interest save to naturalists.

The tongue of a right whale is a soft, thick mass, and has been known to yield twenty-five barrels of oil. Whale flesh is firm, coarse, and red in color.

It has been calculated from the transverse lines on the plates of baleen, each line being supposed to denote an annual check of growth, that whales attain to the age of eight or nine hundred years, but it is by no means certain that this assumption is well grounded.

Part V—Excerpts from *The Gam*

The infant whale is from eight to twelve feet long, and can swim the moment it is born. The mother shows every mark of fondness for her offspring, and the little one, in itself of slight value, is sometimes harpooned and drawn alongside a boat that the mother may be induced to follow. Suckling is done at the surface, mother and baby rolling from side to side that each may breathe in turn.

There is no essential difference in the manner in which the most highly civilized people and the rudest tribes prosecute whale fishery. Both approach the animal in boats, and attack it by harpoons to which lines are affixed, following up and repeating the attack until the strength of the whale is exhausted and it is obliged to succumb.

The most simple harpoon used in whale fishery is a spear about three feet long, with a flattened point, which has sharp edges and two large flattened barbs. These harpoons are attached to long lines.

Many improvements have been made in whale fishery tools. As much depends upon the bladelike edges of the harpoon's barbs as their power to hold when in. Many ingenious devices of movable barbs have been contrived which close on the shaft of the instrument when entering the animal's side, and open outward as soon as there is any strain on the shaft. Another modern device is the gun harpoon, a short bar of iron with a barb at the end, and a ring with a chain for attachment to the line. This is fired from a gun in the hands of the officer. A very effective expedient was suggested by the eminent toxicologist, Sir R. Christison, of Edinburgh University. By this device glass tubes containing prussic acid are so placed in the shafts of the harpoon that the instant the line is pulled tight they are broken, the poison occasioning instant death. Another mode of employing prussic acid is to enclose a glass tube containing it in a hollow bullet about four inches long, which is fired from a rifle made for the purpose, the bullet also containing an explosive connected with a fuse which is kindled as the rifle is fired, so that the bullet bursts immediately after entering the body of the whale, and spreads its deadly contents through the flesh. Strychnia is sometimes used instead of prussic acid with similar results. A whale killed by these methods only disappears for about five minutes, then comes to the surface and instantly dies. But rapid and effective as are these last-named methods, their use is strongly disliked by whalers, who have an unconquerable aversion to handling the carcass of an animal which has been killed by such deadly poisons, and they are now wholly discontinued.

Whaling is a very ancient industry, as in the ninth century the

Whale-Land and Its Customs

Norwegians sent their quaintly-fashioned boats to Greenland in search of them.

Other early maritime nations showed great spirit in whale fishery, but so vigorously did the Dutch pursue the industry, that in the latter part of the seventeenth century they furnished nearly all Europe with oil.

The New England colonies entered upon the enterprise at a very early date; at first by simply going out in boats on their own shores. When in the eighteenth century these shores had become deserted by whales, ships were fitted out for the northern seas, New Bedford becoming the most important whaling port in the world.

But this industry suffered during the Revolution, the War of 1812, the Civil War, and the disasters of the sea, for thirty-four vessels from New Bedford were wrecked at one time in the north Pacific. But it was ruined at last by the substitution of petroleum for whale oil. The ship owners had to look for other ways of using their great vessels. One of these was quite unique. In the war of 1861, the United States Government, wishing to block the ports of Savannah and Charleston and several less important entries to southern harbors, decided upon the novel plan of sinking a large number of abandoned, stone-laden vessels in these ports. The whaler, because of its peculiar model, was especially adapted to this purpose, and twenty-four of the forty-five crafts which formed the Stone Fleet were bought and refitted at New Bedford. Fifty cents a ton was offered for stone, and many a wall was torn down, and many a cobblestone heap and roadside accumulation leveled to furnish forth the necessary weightings.

The whalers were from three hundred to four hundred tons burthen, and were always built of the best material. A fair price for such a ship was about $22,000, and her three years' outfit cost about $20,000 more. Besides all the necessary provisions for officers and men, there were added casks for the oil, spare sails, cordage, and boats.

Such a vessel carried about thirty-two men.

America is at present more actively engaged in whale fishery than any other country, San Francisco being one of its most important ports. The whale is now sought almost exclusively for the bone, for which no adequate substitute has been discovered, and which, owing to its growing scarcity, commands from three to six dollars a pound.

The Frozen North

But the Captain's stories were not always of the sunny South, with its islands covered with graceful palms and luxuriant verdure, and teeming with life. One tale was of the terrible frozen North, which to know is almost like death, so fearful is its breath, and so fatal the clutch of its awful fingers.

For there the sailors see ice everywhere,—clinging to the spars, and jamming with tremendous solidity about the sides of the vessels, holding them with the grip of death in this land of desolation.

The ships of which the Captain told this story were so staunch and goodly and gracefully named withal. Out from the Golden Gate, from New London's harbor, from the wharf at New Bedford, from the port of Boston, and the shores of far Hawaii had sailed the *Onward*, the *Florence*, the *Clara Bell*, the *Acors Barnes*, the *Josephine*, the *Camilla*, the *St. George*, the *Mount Wollaston*, the *Cornelius Howland*, the *James Allen*, the *Java*, the *Rainbow*, the *Arctic*, the *Desmond*, and the *Three Brothers*. Whalers all, ten from New Bedford, manned by stalwart, brave-hearted fellows, eager for adventure, ripe for risks, alert and hardy, not all speaking one language, but all stimulated by the one hope of a prosperous voyage, and rich remuneration as its results.

It was in 1876, after a successful cruise of seven months, the *Arctic* was crushed in the ice July 7th, off Sea Horse Island, eighteen miles from the Bend, her crew being distributed among the other vessels. On August 1st the fleet reached Point Barrows, where it became completely hemmed in by ice. The *Florence* saved herself by managing to keep in the rear of a grounded iceberg, and the *Rainbow* and *Three Brothers* reached a point of safety at Point Barrows. The rudder of the *Clara Bell* becoming broken, she drifted ashore, and was jammed into the ice, while the other vessels were driven northward by the floating ice, struggling in vain to reach open water. Early in September, they found themselves completely ice-bound off Smith's Bay, twenty or thirty miles from land, and with no prospect of release. Their only hope seemed to lie in the abandonment of the vessels; a course which was finally agreed upon.

Tents made from the sails, rations of bread and meat for twenty-five days, with a change of clothing and a blanket for each man, were stored in the boats, which were to be dragged over the ice. It was hoped that enough open water might be found under the ground ice to float the boats southward till the two ships, supposed to be outside the ice pack, could be found. The men carried the baggage ahead for half a mile, then leaving it, returned to drag the boats forward. The exceedingly rough ice was in many places brittle, and some of the men fell through it, thus becoming drenched, and suffering horribly with cold. The company lay down at night worn, spent, famished with cold, only four miles from its starting point.

The next morning a blinding northeast snow storm was raging, and a number of the wayfarers, sick, lame, and discouraged, turned back to the ships.

On September 6th, the intrepid wanderers reached open water, in which they floated their boats towards the land. On the 9th they sighted the *Rainbow* and *Three Brothers* at Point Barrows, and reached them before night. These vessels were in the vise-like clutches of the ice, and could not move. After a consultation between their crews and the newcomers, it was decided that the whole company should start towards the open sea, probably a hundred and thirty miles away, dragging the boats on sleds across the ice. The sleds were made and the journey begun. At Cape Smith the *Florence* was found, and it was decided that the almost hopeless journey towards the open sea should be abandoned, and that they all should winter on the *Florence* at Cape Barrows. The boats were prepared for whaling, whale meat being the only obtainable food.

On September 13th, there swept out from the east a wind, which must have seemed to the beleaguered ones the very breath of God, soft, warm and continuous, before which the ice broke and parted and floated away; and the released *Florence* set her joyful sails, and turned her prow in the direction of a more genial clime.

On the afternoon of the 18th, the *Rainbow* and the *Three Brothers* joined the *Florence*, bringing the crew of the *Clara Bell*, which had been frozen into the ice. The ships appointed a rendezvous at St. Lawrence Bay, where they would take water. All arrived on the 23rd, and there parted ways, the *Florence* heading for San Francisco, where she arrived later, the others departing for Honolulu.

Of those who remained upon and returned to the vessels, nothing was ever heard. They were probably carried to the northeast by the immense ice packs, which closed them round for ten miles, and there perished.

The loss of property through this fleet was estimated at four hundred and forty-two thousand dollars.

It is, of course, first of human life and human suffering that one thinks in connection with such disasters as this, but there is something very like tragedy in the penniless, disheveled, beaten, bruised, scarcely-saved condition of these men, whose sustenance, and, perhaps, that of wives and children, depends upon the success of their voyages.

On November 5, 1871, there appeared at San Francisco, from Honolulu and Australia, the steamer *Moses Taylor*, reporting the loss of thirty-three whalers, some of them crushed and broken by bergs and floes, all of them abandoned in the Arctic seas. No lives were lost, but the estimated financial loss was one and a half million dollars. Were one to follow the result of that loss into individual lives, numberless tragedies might be brought to light, many new dirges of sorrow sounded.

Whales Has Feelins

"Whales has feelins as well as anybody. They don't like to be stuck in the gizzards, an hauled alongside, an cut in, an tried out in these here boilers no more 'n I do!"
—*Barzy Mack's Biology.*

The whale having gone down, we waited for him to come up again. Three boats danced idly upon the warm Madagascan water—the mate's, the second mate's, and my own. The sun blazed viciously down from a cloudless sky.

We lay well apart, covering a large area of swelling, billowy sea. When the whale came up again, the real battle would begin.

A whole hour we waited in anxious expectance. As is natural at such times, my thoughts, meanwhile, ran back years and years to other whales and other fights. Once when I was a cabin boy I had stood three hours in the stern of a stoven boat, sunk just to the gunwale, while two wounded whales were cutting about and making the water white with their huge flukes, so near that it seemed they must kill me. Was the monster, down below in those vague amethystine depths, preparing some such terrors for the present occasion? I recalled, too, how once a dying whale had brought his spout-hole up against our boat and belched barrels full of gore all over us, so that when I opened my eyes every man was painted red—completely covered with fresh, hot blood, so that we all jumped into the water for a hasty bath. Was this sunken leviathan making ready to serve us thus today? I also remembered how a gigantic spouter had tossed me on his flukes—boat, boat's crew, craft and all—whist!—twenty feet into the air, till it seemed that we'd never comedown; and how I found myself at last launched adrift, clinging to a piece of the steering-oar, which had snapped off at the stern-post of the shattered boat. Had not *this* whale flukes also? How would he use them? Should *I* be his victim? or the mate? or the third mate?

Part V—Excerpts from *The Gam*

There is something delicious in this exciting uncertainty. It makes your blood tingle. It makes your nerves thrill. It makes you feel yourself ready to face the whole world of perils and proudly conquer them all. You stand in the stern-sheets, leaning on the steering-oar, and as you look into the faces of those five stalwart men on the thwarts before you, you tell yourself they are fine heroes, every man Jack of them. Yes, heroes! Soldiers face no greater perils. Soldiers win no worthier laurels. Back of every trophy of military valor, you must needs see human bloodshed, human bereavement, human cruelty, and, far too often, the human lust for name and place. But the whaleman's glories are sullied by no such shameful pollution. Is he rich when his sea-toiling days are done? He has impoverished no one. Instead, he has added to the world's wealth. Is he successful in the pursuit of his calling? No widow and no tearful orphans mourn over his triumphs. Is he proud of his profession? He can claim for it the good name of an honest livelihood, a lawful and law-abiding business, a field for soldierly courage purged of soldierly brutality. Nor do tyranny and oppression follow in his paths. Instead, come only the blessings of a peaceful prosperity.

So, as I was saying, you stand and wait, a-tingle with enthusiasm. You are in your glory now. You would not for the whole world be any other thing but a whaleman. You are glad that your boyhood anticipated this splendid life of adventure, and aspired after its high responsibilities. To its toils and its perils you willingly devote your youth and best manhood. You will be proud, in long years to come, to recount the history of your daring sea-battles.

Few landsmen can understand these things. You must go a-blubber-hunting on your own account, fully to grasp their meaning. In fact I know of only one land-lubber who ever really caught the spirit of the whale-hunt, and that is old Walt Whitman, who wrote those splendid, pictorial lines (albeit they go devoid of rhyme, and, in place of precise metre, have only a feeble and slovenly wobble).

> O the whaleman's joys! O I cruise my old cruise again!
> I feel the ship's motion under me, I feel the Atlantic breezes fanning me.
> I hear the cry again sent down from the mast-head, "*There—she blows!*"
> Again I spring up the rigging to look with the rest—we descend, wild with excitement,
> I leap in the lowered boat, we row toward our prey where he lies,
> We approach stealthy and silent, I see the mountainous mass, lethargic, basking,

I see the harpooner standing up, I see the weapon dart from his vigorous arm;
O swift again far out in the ocean the wounded whale, settling, running to windward, tows me,
Again I see him rise to breathe, we row close again,
I see a lance driven through his side, pressed deep, turned in the wound,
Again we back off, I see him settle again, the life is leaving him fast,
As he rises he spouts blood, I see him swim in circles narrower and narrower, swiftly cutting the water—
I see him die,
He gives one convulsive leap in the centre of the circle, and then falls flat and still in the bloody foam."

Barring the single sentence "I see the mountainous mass," (apparently Whitman thought a whale cruised around two-thirds out of water, like a steam-boat) that is a perfect description of the taking of the whale. It is more than that. It is the picture of the inner experience of the chase and the fight—the joy of it, the glow of it, the wild, fierce thrill of it!

I was wishing with all my heart, as we waited for that submarine lounger to return to the surface, that I could somehow tell which boat would get a chance to fasten to him.

But a sudden end to reminiscence and philosophizing. Look! There is frantic excitement in the mate's boat off to leeward—"Stand up and *give* it to him! Quick, quick, *quick*! See!—a figure erect in the boat's bow—a long shaft wielded in both hands high over the man's head—a momentary poise—a swift, springing motion—a sudden recoil—the harpoon hurtling through the air—the slender line singing after it—the weapon sunk fast in something, and that something sinking rapidly into the depths, dragging the line through the chocks so fast the druggs could do nothing to steady it—fifty fathoms—a hundred—two hundred! The mate and the harpooner have changed places. The men dodged the flying line.

Now followed a fresh period of suspense—anxious, but brief.

After a few minutes, there was a sudden uproar in the second mate's boat. Again the excited cry, "Stand up *quick—give* it to him!" Again a heavy harpoon was sent a-whizzing through the air, and plunging deep into that awful, water-hidden something. Again the confusion in the boat and the preparation for lancing.

Responsive to the stab of this second harpoon, the monster sullenly settled under water. The battle was now well joined. What next?

Suddenly and all unexpected, the whale came up again like a submarine boat. He bumped his back against the blades of the first mate's oars, His shiny black hump stood fully a foot out of water. The men could, feel the damp heat of his spout. We could hear the sound of it.

This time old Blubber had gore in his eye. He was in for carnage and calamity and consternation. He lifted his huge square nose ten feet into the air, and dropping his long under-jaw, deliberately calculated his distance. Then with a hideous swing of his whole appalling mass, he veered round and took that whale-boat into his mouth. His ivory teeth smashed through the cedar clinkerwork. The boat went to pieces like an egg-shell.

The mate's crew flung themselves headlong into the water, and escaped by the skin of their teeth.

Now the whale turned suddenly about. His rage redoubled. He would have blood or die for it. Making for the third officer's boat, he threw his cruel jaw across it, turning it bottom-up and staving it in. Again, as by a miracle, every man escaped unhurt.

A pretty situation! There were now two boats' crews floundering and sputtering in the water, while the whale was lashing the waves into froth with his flukes and sending the suds flying in every direction.

With the one remaining boat, I succeeded in picking up the swimmers, and in ferrying them away to the ship. Fortunately we had not far to travel.

How beautiful the *Clara Bell* looked as the boat came round so that I faced her again! Never had I thought her graceful, half-clipper lines such an exquisitely perfect model. Never had winged-dragon figure-head impressed me as such a consummate triumph of the wood-carver's art. Never had the two white streaks along her side from stem to stern seemed such a splendid decoration. Never, in all the days I had sailed in her, had she looked the white-robed angel-guardian she did now. She stood with her main-yard hauled aback. She nodded and dipped, and rose jauntily on the ocean swell. She was the joyfullest ship on all the seas. We were going back to her.

"Oh, ain't I all-fired glad we got done with that wild-eyed monster?" The speaker was dripping with brine. "I calc'lated I was clean daown-swallered like old Jonah—all-fired sure I were!"

"Maybe I ain't glad, too! oh, maybe not! I could look way down in the dratted brute's dratted big gizzards. Deep? Maybe not. Oh, no. Felt like I was dangled over the drattedest deep pit in the whole dratted world."

"*I* wouldn't touch that there man-eater with the far-end of a spare yard—not for *money*;" said a third, squeezing the salt-water out of his beard,

Whales Has Feelins

"no, not for money. I tell you, mess-mates, it's homicide an' man-slaughter, an' bloody murder with malice aforethought to take an' dump two boats' crews down the gullet o' that pesky man-eater, *I* tell you!"

But think not, gentle reader, that these words were spoken in anything graver than jest.

That this whale was a tough one, I have no inclination to deny—not the slightest. I followed the sea forty-one years, I was captain of a ship twenty-eight years, I have sailed more than a million miles, and I have had a hand in the taking of about twelve thousand barrels of oil; but this fighting leviathan off Fort Dauphin was one of the fiercest bits of blubber I ever raised out of the ocean. Yet neither I nor my men had any thought of surrender. We had already wasted two boats on him and we meant to be paid for the outlay. We insisted upon exacting a war indemnity, payable in sperm oil—a hundred or a hundred and twenty barrels—the more the merrier!

We clambered up the ship's side and over the rail. From the deck we could get a startling view of the enemy. The infuriated beast lay wounded, only a short distance from the ship, thrashing around amongst the floating *debris*—oars, paddles, lanterns, and water-kegs—to say nothing of what remained of our two boats, the one a stoven wreck, the other smashed to splinters. It would have turned a landsman cold and stiff with horror. Nor am I certain that all our crew were anxious to renew the battle.

Be that as it may, we immediately got down two new spare boats from the skids overhead. We made them ready for a desperate encounter as we were going into the face of death. We meant to have the chances in our favor. The boats must be made as light as possible, so as to be able to dart away in an instant, if necessary, when the whale showed fight again. To this end, we hastily prepared the most severely abridged outfit not an inch of line, and not a single piece of craft beyond a gun and a bomb-lance in the mate s boat and two hand-lances in mine. We manned the boats with strong crews. The mate took the second mate along with him, and I took the third mate. We lowered the boats and sped away toward our prey.

How bright, how amber-hued, the southern sunlight, as it fell languorous and beautiful upon the ocean billows! How buoyant the dance of our hurrying boats! How impressive the swoop and the soaring of the white gulls and albatrosses! And yet, I dare say, not one of us responded to the fine romanticism of nature—we were bent upon too desperate an errand. There may be a perennially fascinating charm in whaling life when viewed from afar, but there are times when the business assumes a grim ugliness at close range. The poetry of the sea has always been written by landsmen. It always will be.

Charm assuredly there had been in the suspense and expectant anxiety preceding this desperate fight—charm enough, when its horrors were only a possibility; but *now*, when the mystery was pierced, the terrors become hideous facts, and the nature of the foe fully known, the fun was gone altogether. When a fighting whale has chewed up two of your boats and beaten you roundly in his first pitched battle, it is a little unpleasant to go at him again.

Our blood ran high, as we approached the infuriated monster. His spout stood up as tall as ever. He had been no whit enfeebled by his tremendous exertions. Two harpoons stuck out of his back. His flukes swung in air with deadly force and rapidity.

The mate went to leeward of him and fired a bomb-lance into him, but missed his vitals.

Instantly the wounded creature turned about, heaved his head way out of water, opened his cavernous mouth, and made a frightful lunge for the mate's boat. I was just in time. I stood in the bow of my boat, hardly able to wait long enough to choose the right spot for the stab. I was mad with excitement. I plunged the long lance deep into the whale's vitals, and the blood came belching out of his spout-hole rich and red and warm, and after a few moments our victim turned up dead and in a few moments more we had him in the fluke chains along side the *Clara Bell*.

Deafening indeed were the cheers from the ship's deck when we had won that desperate fight; warm was the hand-grip of mess-mates as we climbed aboard; broad and bland the smile on every sun-browned face! We were all alive. We were all unhurt. We had killed the whale.

The inevitable well-worn joke now went the rounds. "Better have paid your wash-woman!"

"*You* needn't talk, Jack; you're as wet as a draownded shark."

"Don't care if I be. Ain't no gearin' 'tween wash-tubs an' whale-boats. Who said there was?"

"*You* did, you slushy hypocritter, you. Ef 'taint so, then what'd you say I cheated my wash-woman for, jest on accaount o' me bein' in a stoven boat, you loony beach-comber?"

"Clew down your jawin'-tackle, sonnywax! Cheerily, oh!"

This sort of mock-malice stood for the best of good-will. The more those men berated each other the better they felt all around.

Whales Has Feelins

When they took the falls to the windlass and manned the bars it was a joy to hear them sing. Sailor-songs are not metrically faultless, any more than Whitman's poems; but they have the Jack Tar spirit of the forecastle breathing all through them, and here and there a touch of easy humor. This particular song ran, as I remember it, something after this fashion:—

> "O, Johnny was no sailor,
> (Renso, boys, Renso.)
> Still he shipped on a Yankee whaler,
> (Renso, boys, Renso.)
> He could not do his duty,
> (Renso, boys, Renso.)
> And he tried to run away then,
> (Renso, boys, Renso.)
> They caught and brought him back again,
> (Renso, boys, Renso.)
> And he said he never would go again,
> (Renso, boys, Renso.)
> They put him pounding cable,
> (Renso, boys, Renso.)
> And found him very able,
> (Renso, boys, Renso.)
> He said he'd run away no more,
> (Renso, boys, Renso.)
> He only waited to get on shore,
> (Renso, boys, Renso.)
> So when he put his feet on shore,
> (Renso, boys, Renso.)
> A-whaling he would go no more,
> (Renso, boys, Renso.)"

What a whale that was! He was the biggest fellow I ever fell in with. He measured sixty-four feet over all, and he had a sixteen-foot jaw. His flukes stretched sixteen feet from tip to tip. He made a hundred and thirty barrels of oil.

"Think what that old spouter must have weighed," said the mate, when we had got him coopered. "One hundred and thirty barrels at eight pounds a gallon—that makes—let me see—that makes" (scratching of head, squirming of eyebrows, smile of relief at last) "that makes two thousand, seven hundred and sixty pounds of oil."

"Here, here!" I said, "work that out on paper, Mr. Wilson; let's be accurate. I'd really like to get at the facts."

So Mr. Dorman figured it out in unimpeachable black and white. He was right. Thirty-two thousand, seven hundred and sixty pounds of oil!

"Now, Cap'n Robbins," he continued, "you'll grant it's within limits to say, one-third oil, two-thirds waste?"

"Yes, a fair estimate; nobody can dispute that."

The Captain

The Captain does not always talk in the jargon of the fo'c's'le. In fact it might be said he uses it merely when he "spins a yarn" in order to be more realistic. He talks with his family, his friends at home and his townspeople like any well-educated man whose school-days are far behind him but who has learned more of men and things from cruising about the world than any books could teach. So this last chapter shall be told in the every-day language he commonly uses when in port among his fellow-men.

There are several anecdotes of himself that the Captain has forgotten to relate. Perhaps he did not think them of sufficient importance, but I do, and so will you when I have told them as they were told to me.

The Captain, like many another old salt, loves dearly his country's flag, though he does not say much about it. But he has carried it into too many strange countries and welcomed the sight of it like a friend from home in too many foreign ports not to be fond of it.

Though he never went to war he had an opportunity to defend the flag when it was insulted in an alien country.

The Captain used to tell us this story. He said:

"I was seated on day beneath the shade of some great trees in front of an hotel in St. Helena with two other American captains.

"It was very hot and the streets were almost deserted. Nobody seemed to be about. We sat there quietly enjoying ourselves, when we saw three burly-looking sailors coming up the street.

"The belonged on a large English ship that had just anchored in the harbor. They were rough-looking, and evidently meant to make trouble for somebody if they could.

The United States consul's office was exactly opposite us, and of course our flag was flying from a tall staff in front.

"When the sailors came up to it, they began to call out in derision, insulting the stars and stripes. Then they cast off the halyards and hauled it down, cursing in the vilest language the 'bloody Yankee flag,' as they called it, and wrapping it about themselves, trailing it in the dust.

"'We could stand it no longer,' said the Captain. 'We felt it was time we took a hand. So when they began to pitch into the consul's clerk, who came out to try to rescue the flag, we, too laid hold of them and a general fight ensued.

"'We did not intend to hurt the men, but we did mean to hold them until the police came.

"'I had my man down and was holding him with both hands when he reached up and grabbed my long whiskers. He had me then completely at his mercy. I could not release myself.

"'The clerk ran out of the office to relieve me, and in trying to strike down the hands of the man beneath me he gave me a severe blow over my eye with an ebony ruler. It was so sore I had to stay in my room for several days.

"'The sailors were arrested and fined three pounds each. Their captain paid their fine and the police put them on board their ship.

"'We found out afterwards that they came on shore with the intention of making a row and getting shut up, hoping their captain would go without them; but you see their plan did not succeed.'"

It was during the war of 1863 that the Captain was at St. Helena, and he was one day on shore dining at the hotel. There were a number of ladies and gentlemen in the party. Several of the latter were captains of American ships that lay in port.

The Captain said:

"A large English ship had just anchored in the harbor, and her Captain came ashore to take dinner.

"He evidently had left his good manners aboard ship, for he entered the room in a blustering manner, ignoring the ladies present, and seated himself near the head of the table and began to talk to an English officer who was one of the guests, about our Civil War, asking questions in a most offensive manner.

"We were all feeling very much pleased over the news of victory we had received and some one began to tell the captain of Sherman's march to the sea.

"The English captain, whom we afterwards learned was one of England s naval reserves, spoke up in a loud and boasting tone, saying,

'Well, sir, we will go over and help the Southerners whip the Yankees when we get back to England.'

"Captain Kelley, an American and a man of small statue, was sitting near me, and I noticed his temper was rising. He could not sit still in his chair. As the English captain went on with his boisterous and blustering talk, he jumped to his feet, pushed up his coat-sleeves, looked the Englishman full in the face, and said, 'It is not necessary for you to wait to whip Yankees. Come out into the street and I will give you a chance, for I'm a Yankee.'

"Kelley was about one-half the size of the Englishman, who looked thoroughly ashamed. He made a lame sort of an apology and left the house. He got his supplies on board that very afternoon and sailed that night, so we never saw him again."

"How did I happen to be in St. Helena?" said the Captain.

"It happened this way. We lay in the harbor of Mauritius two months repairing damages, and getting new masts, rigging and sails; but we couldn't get a whaling boat.

"My crew were deserting and good men were hard to obtain in that quarter, so as soon as my sails were ready, I put to sea and finished rigging her. If any whales had come in sight we couldn't have taken them, as we had only one boat.

"After being a month at sea, we spoke the *Plover*, another whaler, and got one old boat from her. So, having now two boats, we succeeded in capturing two whales.

"At the end of six months we put into Port Louis and found four new whale-boats had been sent us. But we were in as bad a condition as before, for now we had plenty of boats and no crew to man them.

"We went to the Seychelle Islands and there I shipped nine men. We cruised in the Indian Ocean for two years, with poor success, so I decided to head the bark homeward. We had fine weather round the Cape of Good Hope and steered straight for the Island of St. Helena, where we stayed two weeks, so I had a chance to see all there was in that noted place."

St. Helena lies in the track of all vessels bound from Cape of Good Hope to the United States.

"You know, perhaps, that the island is twenty-eight miles in circumference, and is in latitude 15° 55" South, longitude 5° 42" West. It rises up out of the sea like a great tower on the horizon. You can see it forty miles away, a great blur in the distance, on a clear day. As the vessel approaches it, Dana's Peak, 2,700 feet high, is first seen above the clouds.

Part V—Excerpts from *The Gam*

"The island has good anchorage on the north side abreast of Jamestown.

"I took my family on shore and we visited many parts of the island and all the places that are of world-wide interest on account of their connection with the great Napoleon. The Briars, where he lived while Longwood Old House was being prepared for him, was occupied by my esteemed friend, George Moss, Esq. It is situated on a plateau at the foot of the hills and has a fine garden. It is surrounded by wild and rocky scenery. The Briars has always been kept as it was when Napoleon lived there. In one of the rooms it is said that he gave the dictation to 'Las Casas.'

"Longwood Old House was originally a farm house, but verandas were added and the place otherwise improved. After Napoleon's death the house fell into a state of dilapidation.

"Longwood New House was built for Napoleon, but he never lived in it. It is a one-storied building, and has fifty-six rooms of various sizes.

"It is pleasantly situated in the Eastern part of the island, 1,760 feet above the sea.

"Another interesting building is St. Paul's Church, in whose graveyard are buried many strangers who have died on the island.

"Jamestown is in a valley between two lofty hills and is a picturesque spot, with roads winding up the hills on each side.

"Ladder Hill is 600 feet high, and is crowned with a strong fort with barracks for a regiment of soldiers.

"If one wishes to ascend it he may go by the road which winds up the sides of the hill, or, curiously enough, by a long ladder with 365 steps, which reaches from the town at the bottom to the fortress at the top."

Tell us some of your adventures while there, Captain.

"Yes, I think I have given you geography and history enough. But it is pleasant for me to remember I have been in the very rooms where the great Napoleon planned and thought and regretted his life away.

"Our adventures? Well, here's one. The very first day we went ashore we had dinner in Smith's hotel. While we were eating, down came the walls over our heads covering us with plaster.

"No, it wasn't an earthquake, but the work of the white ant which eats up the woodwork of the houses and down they fall when least expected. I tell you, it is dangerous. So some of the dwellings are made of teak wood and the warehouses of iron, both indestructible by this pest.

"We had many drives about the island, visiting Napoleon's grave and drinking from the spring near Longwood, where he walked every day as long as he was able. It is a pleasant spot, cool and shady from the overhanging willow trees.

"Pleasant as was our stay in St. Helena, after our long ocean voyage, we were glad to point the bow of our vessel homeward.

"Yet we were troubled greatly by rumors we had heard of the Confederate cruisers, always ready to pick up vessels belonging to the North, and we kept a good watch out, I assure you, for such cruisers.

"North of the Bermudas we saw traces of the foe in the shape of an abandoned hull, her masts gone, and bearing the marks of fire. We ran near enough to hail her, but no one was on board.

"At last we, too, ran into danger. When only fifty miles south of Nantucket Shoals, our lookout sighted a steamer two points off our lee bow. I went aloft, fearing the worst. My fears were confirmed when I made her out to be a long, rakish, bark-rigged steamer, standing across our bows and heading towards the north.

"I was sure we were lost. But we had suffered all kinds of perils and were to be spared this. For a large ship had been in sight of us all day about eight miles to the windward, steering in the same direction that we were.

"She was a richer prize than we would be, and we saw through our glasses the steamer overhaul her and send a boat to board her.

"It was about sunset when this happened. So we clapped on all sail, put out our lights and sailed away, fearing lest the steamer should take us; but the next day it was thick and foggy, and we saw the steamer no more.

"About thirty miles south of No Man's Land we fell in with a fleet of fishing vessels and gave them a great fright, for they thought we were the enemy.

"When we lowered a boat to go alongside of them, we could see them pulling in their lines to try to escape us.

"But we soon convinced them we were all right, and exchanged some of our salt pork for a fine mess of fresh mackerel. They told us how the land bore, and the next morning, June 25, at five o' clock, we came to anchor off Butler's flat in the lower harbor, New Bedford.

"We had left St. Helena on the first of May, and it was just three years and eleven months since we had sailed from home on what proved a long, disastrous and unfortunate voyage.

"We had suffered from a severe hurricane, losing our boats, which were not replaced for nearly a year.

"Our officers and crew had deserted us, which completely spoiled our plans.

"Yet our ship was new and well-provisioned, and we were so far from home we could not return.

"We had seen whales enough to overload us with oil, but we could not capture them because we had not men enough to do it.

"Yet, when we got home, oil was so high that what we had brought a good price, so our voyage was a paying one, after all.

"Oh, I must tell you that we found out after we got ashore, the vessel that we saw captured was the Isaac Webb of New York, and that after catching her, the steamer tried to find us, but thanks to the fog and our good fortune, she was unsuccessful."

Index

Ada Hall (ship) 68
Africa 45, 46, 88
Ajack (ship) 52
Alabama (warship) 84
Alaska 2
Azores 77, 83, 84, 86, 97, 125, 126

Bahamas 13
Baltimore 27, 100
Belize 31, 33, 35, 36, 38, 40, 41, 42
Beverly, Massachusetts 5
Black, Jake 18–20
blubber 134, 172, 175, 192
Bonaca 31, 42, 48, 49
Boston 5, 6, 7, 9, 12, 17, 53, 62, 68, 69, 78, 82, 179, 186
British Honduras 5, 31
Browne, J. Ross 3, 99

Caicos Passage 13, 14
Canary Islands 86
Canonicus (warship) 52, 55
Cape Hatteras 68
Cape of Good Hope 199
Cape Verde Islands 6, 45, 129, 144
Capella 87
Cephas 87
Cereta (ship) 42, 43
Charleston 2, 185
Chelsea, Massachusetts 2
Chile 18
City of Austin (ship) 59
City of Waco (ship) 67, 68
City of Wilmington (ship) 55
Civil War 1, 3, 13, 185, 198
Clara Bell (ship) 186, 187, 192, 194
Coggia's Comet 61
Columbus 13, 36, 37

Conception Bay 18
Cozumel 51, 52
Cuba 37, 52, 56

D.A. Small (ship) 86
dePieno, Louis 6
Dictator (warship) 52, 55
Dry Tortugas 52

Edgarton 18
Equator 168, 169
Etchings of a Whaling Cruise 3, 99
Everett, Edward 18, 19

Fayal 71, 81, 82, 84, 124, 125, 127, 129
Fernandez, Juan 18
Flemings, Elie 44
Florida 13, 37, 51, 68
Fort Jefferson 52
Frye, Capt. Joseph 52
Fuerteventura 86, 87

Galveston 58, 66, 67
Garden of Eden 37
Glover's Reef 38
Goliah 30, 34
Gonives Bay 22, 25, 26, 27
Gran Canaria 88
Great Cayman 51
Greenwich 14, 59
Gulf of Mexico 5, 51, 60, 62
Gulf of Paria 36

Haiti 25, 27
Half Moon Key 36, 42
Harbor Island 13
Hawaii 89, 186
Hog Islands 43, 48

Index

Honduras 31, 32, 37, 43, 54
Honolulu 188
Horta 81, 84
Huzzey, Elisha P. 5, 17, 49, 50, 55, 58, 59, 68
Huzzey, Henry 6

Inagua 14
Indian Ocean 129, 199
Ireland 15
Isabella (queen) 36
Isle of May 144
Isle of Sal 144

Jamaica 27, 28
Johnson, Jack 25
Jonah 119, 192

Kelley, Capt. Kelley 199
Kentucky 99, 116, 128
Key Bokell 42
Key West 51, 52, 55, 58, 59, 68
Kidd, Captain 13

Lincoln, Pres. Abraham 52
Long Key 42

Macy, Bob 19, 20
Mann, Bill 141, 150
Maret (ship) 73
Martha's Vineyard 168
Mattapoisett 164
Mauger's Key 33
Mayaguano 14
Melrose, Massachusetts 69
Mermaid (ship) 2
Montauk Point 93

Nantucket 18, 201
Napoleon 200
Nassau 13
Navigator Islands 177
Negril 28
New Bedford 5, 11, 17, 18, 53, 101, 102, 103, 107, 118, 125, 129, 162, 165, 185, 186, 201
New London 89, 90, 93
New Orleans 2, 38, 48, 57
New York 12, 53, 55, 56, 68, 100, 101, 103, 104, 129, 202
New York Guano Company 29
Newport 103, 154

North America (ship) 89, 91
North star 169

Olmsted, Francis Allyn 3
Orinoco River 36

Pacific Ocean 43, 69, 89, 169, 175, 185
Peaks, Captain 17
Pease, Horace 6, 50
Pennsylvania 2, 5, 140
Peru 18, 174
Philadelphia 68, 100, 129
Plover (ship) 199
Polaris 87
Port-au-Prince 13, 14, 15, 22
Port-of-Pines 83
Porto Delgado 74, 75, 77, 79
Porto Helgaville 73
Portsmouth 2, 6, 68
Progresso (ship) 51
Providence 103
Provincetown 2, 3, 69, 86

Reed, Mr. 29, 30
Richards, Capt. 90, 91, 93
Roatan 31, 42, 43, 49, 51
Robbins, Capt. Charles Henry 3, 162, 164
Robbins, Lemuel 164

St. Helena 197, 198, 199, 201
St. Mary Island 76
St. Michael, Azores 73
St. Paul's Church 200
Salt, Capt. Bill 105, 106
San Domingo 13, 14, 15
San Francisco 185, 188
San Salvadore 37
Sand Key Lighthouse 52
Santa Cruz 88
Savannah 185
Sea Horse Island 186
Seychelles Islands 199
Shamokawa Eagle 69
Soper, Mrs. 58, 59
Soper, Capt. Robert 2, 5, 7, 10, 11, 13, 15, 16, 25, 31, 32, 34, 39, 44, 48, 50, 54, 56, 60, 63, 65, 68, 69
Southern Cross 169
Stephen 64, 65, 66
Stone Fleet 185
Streeter, (Anson) 6, 23, 24, 49, 59
Styx (ship) 99, 107, 111, 117, 118, 124

Index

Swan Islands 27, 28, 31, 49
Swift (ship) 164, 165, 166, 167, 170, 171, 173, 175

Tabor 131, 132, 133, 134, 140, 157, 160, 161
Tenirife 87
Tierra del Fuego 170
Titusville 2
Tobey, Capt. Lewis 164
Tortuga 14
Triangle Reef 33
Tucker, Charlie 53

United States 28, 52, 74, 98, 99, 125, 185, 197, 199
Upolu 177, 178
Ursa Major 87
Utila 31, 43

Vermont 6
Virginius 52

Ware (ship) 86
Watkins, George 44
Watling Island 13
West Caicos 14
West Indies 5, 7, 28
Western Islands 98
whale 1, 2, 6, 8, 11, 12, 14, 18, 19, 20, 21, 22, 23, 24, 25, 49, 50, 60, 86, 90, 95, 105, 114, 118, 119, 120, 121, 122, 123, 126, 130, 131, 132, 133, 134, 135, 136, 137, 138, 139, 144, 145, 146, 148, 149, 150, 162, 169, 170, 171, 172, 173, 174, 175, 176, 182, 183, 184, 185, 187, 189, 190, 191, 192, 193, 194, 195, 199
Whitman, Walt 190
William Penn (ship) 178
William Rotch (ship) 170
Williams, Sam 2, 3, 69, 88
Wm. A. Grozier (ship) 86
Worcester (warship) 52, 54, 55

Yucatán 51

www.ingramcontent.com/pod-product-compliance
Ingram Content Group UK Ltd.
Pitfield, Milton Keynes, MK11 3LW, UK
UKHW042003140426
5217IPUK00015B/960